AF411350

PV Charging and Storage for Electric Vehicles

PV Charging and Storage for Electric Vehicles

Editors

Pavol Bauer
Gautham Ram Chandra Mouli

MDPI • Basel • Beijing • Wuhan • Barcelona • Belgrade • Manchester • Tokyo • Cluj • Tianjin

Editors
Pavol Bauer
DC Systems, Energy Conversion
and Storage Group, Department
of Electrical Sustainable Energy,
Delft University of Technology
The Netherlands

Gautham Ram Chandra Mouli
DC Systems, Energy Conversion
and Storage Group, Department
of Electrical Sustainable Energy,
Delft University of Technology
The Netherlands

Editorial Office
MDPI
St. Alban-Anlage 66
4052 Basel, Switzerland

This is a reprint of articles from the Special Issue published online in the open access journal *Energies* (ISSN 1996-1073) (available at: https://www.mdpi.com/journal/energies/special_issues/PV_Charging_and_Storage_of_Electricity_Vehicles).

For citation purposes, cite each article independently as indicated on the article page online and as indicated below:

LastName, A.A.; LastName, B.B.; LastName, C.C. Article Title. *Journal Name* **Year**, *Volume Number*, Page Range.

ISBN 978-3-0365-0104-8 (Hbk)
ISBN 978-3-0365-0105-5 (PDF)

Cover image courtesy of US Department of State (Public Domain).

Contents

About the Editors

Pavol Bauer is currently a full Professor with the Department of Electrical Sustainable Energy of Delft University of Technology and head of DC Systems, Energy Conversion and Storage group. He received a master's degree in Electrical Engineering at the Technical University of Kosice ('85), PhD from Delft University of Technology ('95) and the title 'Prof.' from the President of Czech Republic at the Brno University of Technology (2008) and Delft University of Technology (2016). He is also an honorary professor at Politehnica University Timisioira in Romania. From 2002 to 2003, he was working partially at KEMA (DNV GL, Arnhem) on different projects related to power electronics applications in power systems. He published over 110 journal and over 400 conference papers in his field (with H factor Google scholar 38, Web of Science 25), he is an author or co-author of 8 books, holds 5 international patents and has organized several tutorials at international conferences.

He has worked on many projects for industry concerning wind and wave energy, power electronic applications for power systems such as Smarttrafo; HVDC systems, projects for smart cities such as PV charging of electric vehicles, PV and storage integration, contactless charging; and he participated in several Leonardo da Vinci, H2020 and Electric Mobility Europe EU projects as project partner (ELINA, INETELE, E-Pragmatic, Micact, Trolly 2.0, OSCD) and coordinator (PEMCWebLab.com-Edipe, SustEner, Eranet DCMICRO). He is a Senior Member of the IEEE ('97), former chairman of Benelux IEEE Joint Industry Applications Society, Power Electronics and Power Engineering Society chapter, chairman of the Power Electronics and Motion Control (PEMC) council, member of the Executive Committee of European Power Electronics Association (EPE) and also member of international steering committee at numerous conferences.

Gautham Ram Chandra Mouli is an Assistant Professor in the DC systems, Energy conversion and Storage group in the Department of Electrical Sustainable Energy at the Delft University of Technology, The Netherlands. His current research focuses on electric vehicles, EV charging, PV systems, power electronics and demand-side management.

He received his bachelor's and master's degrees in Electrical Engineering from the National Institute of Technology Trichy, India in 2011 and the Delft University of Technology in 2013, respectively. He received his PhD from Delft University in 2018 for the development of a solar-powered V2G electric vehicle charger and designed smart charging algorithms (with PRE Power Developers, ABB and UT Austin). It was awarded the 'Most significant innovation in electric vehicles' award from IDtechEx in 2018. From 2017 to 2019, he was a postdoctoral researcher at TU Delft, working on research topics related to power converters for EV charging, the smart charging of EVs, and trolley buses.

He is involved in many projects with industrial and academic partners at national and EU levels, concerning electric mobility and renewable energy such as PV charging of electric vehicles, OSCD, Trolley 2.0, Flexgrid. He was awarded the best paper prize in the IEEE Transactions on Industrial Informatics in 2018, the Best Poster prize at Erasmus Energy Forum 2016, Netherlands, and the Best Paper prize at the IEEE INDICON Conference 2009, India. He is the coordinator and a lecturer for Massive Open Online Course (MOOC) on Electric cars on edX.org with 125,000 learners from 175 countries. He is the Vice-chair of IEEE Industrial Electronic Society Benelux chapter.

Preface to "PV Charging and Storage for Electric Vehicles"

Two major trends in energy usage that are expected for the future are the increase in distributed renewable generation like solar energy, and the emergence of electric vehicles (EV) as the future mode of transportation. At the same time, there are many challenges for the integration of these two technologies. Firstly, electric vehicles are only 'green' as long as the source of electricity is 'green' as well. Secondly, photovoltaic (PV) power production suffers from diurnal and seasonal variations, creating the need for energy storage technology. Thirdly, overloading and voltage problems are expected in the distributed network due to the high penetration of distributed generation and increased power demand from the charging of electric vehicles.

The energy and mobility transition calls for novel technological innovations in the field of sustainable electric mobility powered from renewable energy. This Special Issue focuses on recent advances in technology for PV charging and storage for electric vehicles and includes, but is not limited to, the following topics:

- Power electronic converter for (DC) charging of EVs from solar (with bidirectional capability to feed energy back to the grid);

- Investigation of the synergy between solar electricity generation and EV charging demand;

- Innovative design of electric vehicles with on-board solar power for increased driving range;

- Intelligent systems for off-grid (stand-alone) solar charging of EVs;

- Power management techniques for solar EV systems to reduce grid congestion, increase solar self-consumption, reduce energy costs, and increase grid stability;

- Optimal sizing, location, and control of energy storage to manage diurnal and seasonal solar variations in order to meet EV charging requirements;

- Charging electric vehicles from solar energy in microgrids;

- Recent developments in ICT protocols for solar-powered smart charging of EVs (with V2G);

- Novel solar-powered contactless EV charging system (with bidirectional power capability to feed energy back to the grid);

- Solar-powered electrified public transportation (e.g., trams, buses, trains);

- Using the EV as energy storage for PV via Vehicle-to-X (e.g., V2G, V2H, V2B, V2L);

- State-of-the-art reviews on solar charging of EVs.

We sincerely thank the authors, the reviewers and the staff of MDPI for their contributions to this issue.

Pavol Bauer, Gautham Ram Chandra Mouli
Editors

Article

A Model for Cost–Benefit Analysis of Privately Owned Vehicle-to-Grid Solutions

Jesús Rodríguez-Molina [1],*, Pedro Castillejo [1], Victoria Beltran [2] and Margarita Martínez-Núñez [3]

[1]   Department of Telematics and Electronics Engineering, Technical University of Madrid, 28031 Madrid, Spain; pedro.castillejo@upm.es

[2]   Department of Electronics, Computer Technology and Projects, Technical University of Cartagena, 30203 Cartagena, Spain; victoria.beltran@upct.es

[3]   Department of Organization Engineering, Business Administration and Statistics, Technical University of Madrid, 28031 Madrid, Spain; margarita.martinez@upm.es

*   Correspondence: jesus.rodriguezm@upm.es; Tel.: +34–910673350

Received: 20 August 2020; Accepted: 23 October 2020; Published: 6 November 2020

Abstract: Although the increasing adoption of electric vehicles (EVs) is overall positive for the environment and for the sustainable use of resources, the extra effort that requires purchasing an EV when compared to an equivalent internal combustion engine (ICE) competitor make them less appealing from an economical point of view. In addition to that, there are other challenges in EVs (autonomy, battery, recharge time, etc.) that are non-existent in ICE vehicles. Nevertheless, the possibility of providing electricity to the power grid via vehicle-to-grid technology (V2G), along with lower maintenance costs, could prove that EVs are the most economically efficient option in the long run. Indeed, enabling V2G would make EVs capable of saving some costs for their vehicle owners, thus making them a better long-term mobility choice that could trigger deep changes in habits of vehicle owners. This paper describes a cost–benefit analysis of how consumers can make use of V2G solutions, in a way that they can use their vehicle for transport purposes and obtain revenues when injecting energy into the power grid.

Keywords: electric vehicle; vehicle-to-grid; cost–benefit analysis

---

1. Introduction

The smart grid is one of the most promising infrastructures developed during the last years for the improvement of access to electricity and its usage, as it is bringing key benefits: a combination of existing information and communication technology standards [1], the power grid itself to enhance the stability of the system [2,3], and the incorporation of new actors in the energy markets [4]. Among other features, the smart grid enables a set of activities aimed to the demand side management, used to optimize energy usage according to specific characteristics of demand response systems, energy efficiency, or usage time of the resources [5,6]. Other applications such as home load control and home energy management [7] are covered as well. Energy storage is a major feature, due to the fact that it has to be enabled and balanced in distributed-like systems for increased effectiveness [8] and can be used to trade it in the aforementioned energy markets or to provide energy in moments where it cannot be harvested from the environment (like photovoltaic deployments during the night). More importantly, it allows prosumers (that is, energy consumers able to produce their own electricity by means of distributed energy resources) to have more energy available for their private use and utilize the surplus power they produce as a source of revenues. In this regard, electric vehicles (EVs) may become an appealing solution—especially when compared to vehicles with an internal combustion engine (ICE)—as they are capable of having vehicle-to-grid (V2G) characteristics that will enable them to inject electricity into the power grid, resulting in an opportunity to create income for the V2G vehicle's owner. To this end, it is necessary to add some specific infrastructure to the EV, namely, a V2G bidirectional power converter, or to change the software configuration of its electric charger. Furthermore, this V2G approach leads to new trade opportunities that were not possible before, as for example selling energy to an aggregator located between the distributed system operator and the prosumers.

This paper studies how privately owned V2Gs can compete with ICE-based vehicles in terms of economic efficiency, putting forward some scenarios where a new mathematical model has been demonstrated. The cost–benefit model that is presented in this manuscript shows a thorough comparison of the expenses between EVs and ICE vehicles during an extended period of time, as well as an economic assessment between purchasing and renting the battery of an EV and how costs vary depending on several different profiles of vehicle usage. The authors have established a comparison between ICE and V2G vehicles because the objective of the manuscript is to assess if V2G technology can be used to make EVs economically competitive when compared to the traditional ICE-powered automobiles. Typically, and especially if subsidies are removed, the cost of an EV is higher than a comparable ICE vehicle. Even though maintenance costs and electricity are lower than gas and maintenance of ICE-powered vehicles, it is at least arguable whether at the end of its timespan of usage an EV is more economical than an ICE vehicle. Nevertheless, by using V2G technology, an EV should be better positioned to reduce costs in mobility with privately owned vehicles. It is the authors' opinion that it is interesting to have a study on the matter of comparing ICE vehicles with EV-V2G ones, as it could provide a more accurate perspective on how advantageous it is to use V2G technology in an EV to reduce expenses. Considering a set of parameters and how they relate with each other, an analysis of the obtained calculations has been carried out for certain cases. The contributions of this paper are as follows:

1. A thorough mathematical cost–benefit model used to analyse accurately to what extent V2G technology can be profitable for a regular end user. This model takes into account parameters that, to the best of the authors' knowledge, have not been included in any other cost–benefit model that involves different end users for V2G technology depending on their vehicle user profile, like battery discharging while being idle, how the depreciation of the electric vehicle influences its maintenance costs, or the suboptimal trade of electricity that might happen if the vehicle is not available during the most suitable time slots of the day to be recharged.
2. The differences in expenditures for V2G solutions when the battery is purchased with the whole vehicle or leased from the manufacturer. This is another contribution that the authors of this paper have not seen in the existing literature about this topic.
3. The application of the cost–benefit model to three user profiles for V2G and ICE solutions, along with how they fare after a prolonged period of time. The authors believe that this adds a realistic justification with several examples that make use of actual data and parameter values extracted from updated references in order to know to what extent using V2G may provide an economical benefit to their end users.
4. A review of comparable models that have been created by other authors, pointing out the main challenges that have still to be dealt with and why the one put forward by the authors of this manuscript represents an improvement over the previous ones.

This paper is structured as follows: an introduction has already been offered as the first section. Section 2 offers a compilation of related works. Section 3 describes the variables included to elaborate the model. Section 4 presents the model. Section 5 offers the numerical evaluation of the model when facing two different possible scenarios. Section 6 explains the conclusions obtained from the study. Acknowledgments and references are displayed as the final parts of the manuscript.

2. Related Works

The studies done about the possible applicability of a V2G solution for particular environments have been included in this section, along with the open issues that have been found in the reviewed literature.

2.1. State of The Art

L. Noel and R. McCormack put forward their own cost–benefit analysis when comparing a V2G-capable electric school bus with a diesel-powered one [9]. They take into account a large set of variables (including seating capacity, cost of electricity, cost of diesel fuel, etc.), the authors conclude that using a school bus with V2G capabilities is more cost-effective than a diesel one when V2G capabilities are enabled, thus making the latter almost mandatory (savings up to $6070 per seat are claimed). However, the study focuses on municipal school buses (which are more expensive and far less abundant than automobiles), rather than private transportation. This study has been challenged by the one presented by Y. Shirazi et al. [10], where it is mentioned that, as far as Philadelphia and its school district are concerned, a V2G bus is not cost effective and it actually increases its usage costs compared to a diesel-powered one. The reasons behind this conclusion involve limitations that, according to the authors, are inherent to electric vehicles and are often overlooked, such as low environmental temperatures or electrical losses resulting from V2G technology.

D. Park et al. offer a cost–benefit analysis where it is claimed that savings with EV services range from $8000 to $22,000 per year and per vehicle in an optimized frequency regulation (FR) market [11], which is the one that best adapts to the nature of V2G services, due to its pattern of energy supply in bursts rather than as a constant and reliable flow source. The authors consider fine-grained characteristics like daily mobility patterns and mobility model velocities. The study that has been done here, though, only covers municipal services (school transport, waste collecting truck, and city bus) rather than private vehicles.

O. A. Nworgu et al. describe the economic prospects of V2G technology in the electric distribution network [12]. They mention how V2G infrastructure can be used for valley filling during low demand periods and peak shaving when electricity demand is high. However, their model does not take into account the energy losses resulting from using V2G as a way to store and transfer energy (rather than a regular generator or home battery) or the required cost to adapt an EV to V2G technology.

D. M. Hill et al. describe fleet operator risks for V2G regulation [13]. A V2G fleet financial model is displayed where the replacement of ICE trucks with extended range electric vehicles is studied, considering three scenarios where this replacement may or may not be cost efficient. Battery degradation and replacement, which easily comes as one of the most significant challenges of V2G technology, are fully considered, as well as risk acceptance for vehicle owners that might be unwilling to switch to this kind of technology. The authors' proposal, though, is focused on fleets of vehicles rather than private transport.

M. Musio et al. consider the added benefits of having V2G technology working as a virtual power plant (VPP) [14]. The authors stress the importance of having a suitable battery available for this kind of technology and offer a thorough study on a simulation of a battery lifetime in terms of charge and discharge. In addition to that, a case study is displayed where an optimization problem, understood as the number of EVs that minimizes the cost of the VPP, is reasoned. However, the authors explicitly mention that the resulting VPP works autonomously with no trade activities with the main grid, as it has likewise been considered in this manuscript.

P. Jain et al. also mention a similar idea with aggregated EVs included in a V2G-based power service [15]. Different kinds of vehicles are taken into account for the estimations done regarding revenue evaluation. The aggregated electricity provided by the V2G network is assessed as the aggregated state of charge (SOC) of the batteries. However, the work presented by the authors deals with specific information that has been obtained from external sources and they perform the calculations based on them, rather than attempting to offer a new model.

H. Lund and W. Kempton describe in [16] how renewable energies can be integrated in the transport sector via V2G. The authors present a model, referred to as EnergyPLAN, which they have developed under a framework of national level energy devoted for transport, heat and electricity. V2G plays a prominent role in this model, due to the fact that the sharing of vehicles enabled with this technology that is connected to the grid is expected to provide power to the grid. The number of inputs that have been used in the model to define EVs with V2G are fewer than the ones that have been considered in this manuscript, though. The authors have considered the transportation demand of electric cars, share of V2G solutions both being driven during peak hours and connected to the grid, efficiency of the chargers and inverters, capacity of the battery, distribution of the transportation demand, and the power capacity of the grid connection.

H. Qiang et al. put forward a mathematical model [17] where the initial SOC, charging power and initial charging time are assessed with the objective of obtaining a more accurate way to compute the charging load used by private EVs. Their model takes into account the SOC of the battery, the initial SOC of charging and the charging power, but it falls short when considering other features more related to an economical point of view, such as battery degradation, inflation or the battery costs.

Santoshkumar et al. propose an architectural framework of an off-board V2G integrator for the Smart Grid [18]. They refer to off-board integrators as the ones that are outside of vehicles and are able to connect several EVs to the power grid. In the mathematical model that they put forward there are several features that have been taken into account for the testing activities that the authors have carried out: power of the domestic loads efficiency of the chargers or the number of existent EVs are some of them. Unfortunately, the features involved by the scope of this manuscript, which are used to demonstrate the economic feasibility of the integration of V2G technology in the smart grid are not present.

Chenggang Du and Jinghan He also mention how a strategy for multiple V2G solutions can be applied for their batteries' charge and discharge [19]. According to the authors, this charge–discharge plan would be able to lower differences between peak and valley energy demand hours significantly. Among other characteristics, power and energy restrains are taken into account to create the daily load curve that is obtained after enhancing daily energy consumption with the integration of V2G technology. As it happened with some previous proposals, this one models quite accurately features related to electricity and power but does not take into account the potential economic benefits of V2G owners.

Zesen Wang et al. describe in [20] their own contributions to the usage of V2G technology for building-integrated energy systems (referred to as BIES). They determine how vehicles with this technology can be used as movable energy storage devices capable of providing electricity to other loads. V2G plays a supportive role in the suggested model, as simulations have been used to prove that a fleet of V2G equipment can improve the overall economy of BIES. However, the authors of this paper have focused on the role of V2G within a BIES, rather than making a BIES part of the grid or focusing it as a specific solution for end users.

Yuancheng Li et al. show in [21] how differential privacy is an important matter to consider in V2G networks. The overall structure of a V2G is introduced, and the roles of each of its entities (control center, aggregator, distribution network, and charge station) are described as well. As far as privacy protection is concerned, a spatial data decomposition algorithm is put forward by the authors. Experimental results obtained from the charging positions of 100,000 electric vehicles and 1500 public charging posts have been presented. However, the researchers´ main purpose is to address differential privacy in the charging infrastructure of V2G networks, rather than presenting a cost–benefit analysis.

Tohid Harighi et al. make an overview of storage systems, energy scenarios and the required infrastructure for V2G technology [22]. It is regarded as part of the overall infrastructure that would be required to decrease greenhouse gases (GHG) to an acceptable minimum that meets the targets that have been agreed for 2050. Unfortunately, the paper does not offer a mathematical model on how to integrate V2G technology in a larger network, nor it provides a cost benefit analysis on the profit possibilities offered by V2G.

Michael Child et al. estimate in [23] how a significant amount of V2G solutions could impact a completely renewable system. The authors of this manuscript have used the above-mentioned EnergyPLAN modelling tool as a way to assess the impact of the contributions that can be done by a V2G network. A thorough assessment on how energy would be consumed, supplied and stored is made in the manuscript. There is no cost–benefit analysis model presented by the authors, though.

Another study based on comparisons between long-term usage of EVs and ICE vehicles is the one made by Peter Weldon et al. in [24]. The authors show how, under the specific use case of Irish infrastructure and economic incentives to buy EVs, different levels of economic competitiveness of EVs over ICE vehicles can be achieved. The authors have studied four different kinds of comparable vehicles (small, medium, large, and vans) for both kinds of energy sources and have reached several conclusions: after a 10-year period of time, EVs are more economically efficient in almost every possible situation, except when gasoline prices remain constant. Overall, the paper describes the situation that would take place in scenarios where vehicles have high, medium, or low frequency of usage and the conclusions reached are close to the ones that we have obtained as well. However, battery degradation is not considered as detailed as in this manuscript, nor there is information on efficiency with V2G solutions. Lastly, externalities are not taken into account, and battery replacement is only considered for the high frequency usage case, which is to be expected since regular EVs that do not make use of V2G facilities should not require such an action.

A similar study is shown by Yiling Zhang et al. [25]. In this case, V2G has been studied as a technology oriented to car sharing. In order to quantify the potential benefits from using it, a model making use of two-stage stochastic integer program has been considered. An estimation of the benefits of integration has been made by the authors, which includes the benefits that will result from the energy trade, as well as costs related to vehicle relocation and charging. This study, though, is not targeting battery degradation as a major factor as it is done in our manuscript, and there are no different user profiles for the model that has been created.

In the piece of research made by Kyuho Maeng et al. [26] the integration of V2G into the grid and what benefits it can provide are major topics for research as well. The authors of this paper offer a mixed multiple discrete-continuous extreme value (MDCEV) model based on random utility theory (RUM). The model is used to obtain market simulation results that define what kind of vehicle would be preferable for a sample of Korean population. This study, though, does not consider profitability for end users as one core concept, nor battery degradation is taken into account in a thorough manner.

There are also other references that consider externalities for V2G technology. For example, it is shown in [27] that, "BE [Battery Electric] transit and school buses with V2G application have potential to reduce electricity generation related greenhouse-gas emissions by 1067 and 1420 tons of CO2 equivalence (average), and eliminate \$13,000 and \$18,300 air pollution externalities (average), respectively". Air externalities are compared between V2G and ICE (diesel) mobility solutions, along with the V2G technology cost for similar school and transit buses. According to this manuscript, in the CAISO (California ISO) region, V2G makes possible that the lifetime total cost of an electric school bus is little more than a sixth of the cost in the diesel one, whereas costs for a regular transit bus are around a fifth lower for V2G than for the ICE solution. However, as it happened in other cases, the study is not applied to private transport. In addition to that, it is stated in [28] that if externalities are taken into account for generation, new storage (where V2G solutions are included) and new loads to model a large regional transmission organization, 50% of renewable energy should be implemented. This study is more focused on externalities than in V2G usage, though. The usage of V2G combined with other smart grid technologies has been subject of research as well. For example, demand response (DR) is the main focus on [29]. The study proves how using V2G in specific moments such as night time can improve the overall regularity of energy consumption (a feature most looked into from the point of view of the electricity supplier) with the aid of a home energy management (HEM) system, smart meters and V2G itself. It is also mentioned that V2G can put a strain on loads working during the night, as they can increase in number due to low energy prices during that time slot. The interaction between demand response management and V2G is also studied in [30], where it is explained that their cooperation is critical to use surplus energy in EVs to the end user´s advantage. The system that is put forward takes an auction-like approach: by means of having EVs selling electricity under dynamic pricing to a number of aggregators, the latter compete to obtain the best possible price, while at the same time offering incentives to EVs to act as V2G solutions.

Finally, there are some more studies that have researched on the economic and energy charging possibilities of comparable EV and EVV2G solutions. For example, it is claimed in [31] that dynamic EV scheduling charge/discharge can optimize V2G usage and capacity. The authors of the manuscript describe how an algorithm built as part of their building energy management system (BEMS) can be used for 30 min V2G capacity estimations. Their model has been tested for three different use cases (high-rise residential buildings, office buildings, and commercial buildings) and the researchers mention how using several EVs as distributed energy storage can be possible for high-rise buildings. Long-term costs compared with vehicles with ICE-powered vehicles is out of the scope of the manuscript, though. Additionally, it is studied in [32] how different charging schemes with or without the usage of V2G can offer complementary results. The authors discuss four charging modes (night charging, night charging with V2G, 24 h charging, and 24 h charging with V2G) and study how they impact in vehicle usage. It is also mentioned how V2G provides an opportunity to profit through electricity arbitrage by discharging energy to the power grid during non-driving periods of time. This piece of work, however, is focused on the different charging possibilities for an EV rather than its long-term economic performance compared to the one that an ICE vehicle offers. Lastly, a model for communications based on the long term evolution (LTE) protocol among EVs that make use of V2G technology is described in [33]. The researchers claim how this protocol can be used to communicate two EVs wirelessly by making use of the physical layer present in the LTE protocol. In this way, it is claimed that an aggregator can send information to EVs about power requirements on an area under its range, and in case a V2G is unaware of the power demand, the LTE system will send the information from a regular EV to a V2G automobile. State of charge in the battery of the EV is the main feature used to establish whether power will be bought or sold.

The most prominent features from the reviewed literature have been included in Nomenclature. Many of these studies' strong points have been taken into consideration for the mathematical model that is presented in this manuscript. For example, battery degradation and replacement are a major part of the studio that has been done, whereas weaknesses in Table 1 like lack of attention to private transport have been sufficiently covered in the mathematical model put forward in this manuscript.

Table 1. Summarization of the main advantages and disadvantages of the reviewed literature.

Reviewed Work	Strengths	Weaknesses
L. Noel and R. McCormack [9]	Complete cost–benefit analysis for public transport	Focused on school buses rather than private transportation. This manuscript has been challenged by [10]
Y. Shirazi et al. [10]	Provides more parameters to consider (low temperatures, electrical losses)	Focused on school buses rather than private transportation
D. Park et al. [11]	Mobility patterns and mobility model velocities are taken into account	The study only covers municipal services
O. A. Nworgu et al. [12]	It is mentioned how to use V2G to flatten demand curve	The model does not take into account energy losses from using V2G
D. M. Hill et al. [13]	Battery degradation, replacement and risk acceptance are taken into account	The proposal deals with fleets of vehicles rather than private transport
M. Musio et al. [14]	The optimization problem resulting from having vehicles acting as a VPP is analyzed	The resulting VPP has no trade activities with the main grid
P. Jain et al. [15]	Perspective on SOC of the batteries is provided	Calculations are done based on external sources rather than by providing a new model
H. Lund and W. Kempton [16]	Model that integrates energy used for transport, heat and electricity at a national level	Less variables are taken into account than in the model presented in the manuscript

Table 1. *Cont.*

Reviewed Work	Strengths	Weaknesses
H. Qiang et al. [17]	Initial SOC, charging power and initial charging time are considered	Battery degradation, inflation or the battery costs are not considered
Santoshkumar et al. [18]	Varied loads have been taken into account in the model	economic feasibility of V2G integration is not present
Chenggang Du and Jinghan He [19]	Power and energy restrains are used for the daily load curve	The model does not take into account V2G owners
Zesen Wang et al. [20]	V2G solutions are modelled as movable energy storage devices	The model regards V2G as a support for Building-Integrated Energy Systems
Yuancheng Li et al. [21]	A thorough experimental analysis has been done regarding location privacy	The model focuses on differential privacy in V2G rather than doing a cost–benefit analysis
Tohid Harighi et al. [22]	V2G is acknowledged as a technology that can be used to meet goals in GHG reduction	Neither mathematical model nor cost–benefit analysis for V2G are offered
Michael Child et al. [23]	Impact of V2G in a system completely based on renewable energies is assessed	No cost benefit analysis has been performed in the manuscript
Peter Weldon et al. [24]	Model with three different kinds of EV users	No data about V2G solutions or battery degradation. Externalities not considered
Yiling Zhang et al. [25]	Study on the integration of V2G into the electricity grid	No data about battery degradation. No different profiles
Kyuho Maeng et al. [26]	Study of the most preferable kind of vehicle for a significant sample of users	No data about battery degradation. Profitability for end users is not considered

2.2. Open Issues

There are several open issues that required to be tackled if an accurate, objective assessment of V2G technology is going to be done.

1. Limitations in the mathematical models. Despite the efforts done by the authors, the mathematical models used show limitations that make them obsolete after a relatively short amount of time. A model that offers a significant number of parameters to measure accurately the cost–benefit of V2G solutions, while at the same time keeping the time required to perform calculations at a reasonable level, is required.

2. Lack of orientation to private transport. The evaluations that are done in the literature are mostly concerned about transport fleet or public services. However, the adoption of these solutions by private owners of vehicles is a critical point for V2G, as they are more numerous and the cost of their means of transport is lower when compared to a school bus or a truck. While there are literature references proving that EV owners could be interested in enabling their vehicles with V2G technology if given suitable options ("Our findings suggest that the V2G concept is most likely to help EVs on the market if power aggregators operate either on pay-as-you-go basis", [34]) they do not show a mathematical model that takes into account different EV user profiles, externalities or battery usage options.

3. Reduced scope of numerical results. The studied literature usually reflects how a model can be applied or not by taking into account too specific situations, such as public transport in a city or any other location that is very dependent on meteorological circumstances, the kind of public service that is attended or the route that is taken every day by the vehicles, so it becomes difficult to make an objective analysis of those scenarios.

All these challenges have been born in mind to design the mathematical model presented in this manuscript, as well as the calculations and results placed in the next sections. Overall, the model can be described as depicted in Figure 1. The main figures that have been taken into account are acquisition and operational costs, externalities inherent to the vehicle, gas consumption, and maintenance, among others.

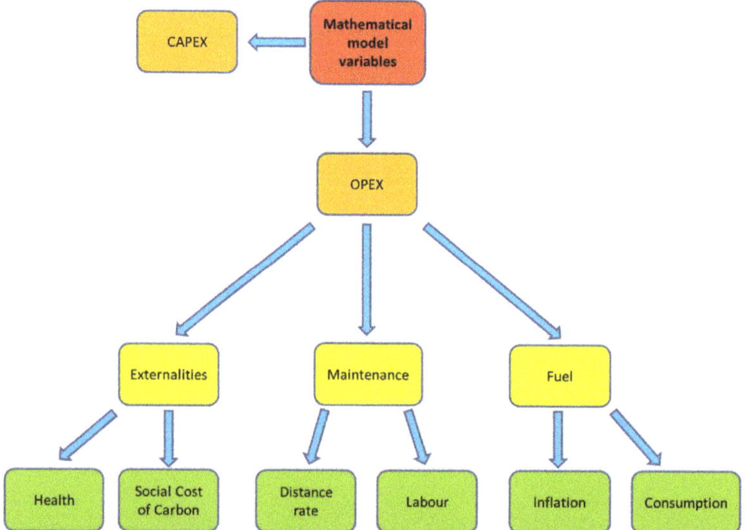

Figure 1. Common variables considered for the mathematical model and their relations.

3. Mathematical Model for V2G Integration

The costs that have to be faced by an individual (or a small group like a family that will use a single automobile) can be defined as C_{totICE} for the ICE vehicle, and C_{totV2G} for the vehicle-to-grid solution, whereas *Capex* figures for the ICE and the V2G solution could represent the cost of acquiring the vehicle by an individual. Finally, *Opex* figures for the ICE and the V2G vehicles represent the mandatory operational expenses needed to have the asset fully functional. Thus, the total costs for each of the transport solutions can be expressed as in (1) and (2):

$$C_{totICE} = Capex_{ICE} + Opex_{ICE} \tag{1}$$

$$C_{totV2G} = Capex_{V2G} + Opex_{V2G} \tag{2}$$

Note that it has been chosen to consider each of the vehicles as an asset rather than a liability due to V2G potential to generate revenues or at least reduce operational costs. Inflation can also be considered in the *Opex* expenditures by adding its corresponding parameter (represented by *Inf*) into the previous equations, as long as it is defined for a specific amount of time. Typically, inflation will build up as time goes by as a function, or a part of one, where time piles up on an exponential basis. Therefore, inflation-adjusted prices have been added as shown in (3) and (4).

$$C_{totICE} = Capex_{ICE} + \sum_{i=1}^{j} (1 + Inf)^i \times Opex_{ICE} \tag{3}$$

$$C_{totV2G} = Capex_{V2G} + \sum_{i=1}^{j} (1 + Inf)^i \times Opex_{V2G} \tag{4}$$

Additionally, *Capex* for the EV must be further defined as the costs of installing the required infrastructure to transform the EV into a vehicle with V2G capabilities (represented as *Vconv*) and to charge the EV at home (represented by *Heq*), along with the cost of the vehicle itself (C_{EV}). It has been done in (5). As it can be inferred, these technological needs do not apply for the internal combustion engine vehicle.

$$Capex_{V2G} = C_{EV} + Vconv + Heq \tag{5}$$

The *Opex* for ICE and EV must be further analyzed. It is shown in (6) what the *Opex* value is for an ICE vehicle. During a period of time that ranges from *i* to *j*—considering *i* as a year and *j* as twelve, which has been estimated as the average lifetime of a vehicle, according to a) what is used in [24] and b) the estimation done in [25]—four aspects will add up to the final figure of the expenses: the yearly costs of the externalities of the vehicle (Ex_{ICE}), fuel consumption (F_{ICE}), and maintenance expenses (M_{ICE}). As far as the V2G solution is concerned the equation is represented by the same kind of terms used for the ICE vehicle (7). Nevertheless, contrary to ICE vehicles, in this case, the expenses, maintenance and electricity costs need to be further defined, as it is detailed through the following subsections.

$$Opex_{ICE} = \sum_{i=1}^{j} (Ex_{ICE} + M_{ICE} + F_{ICE})_i \tag{6}$$

$$Opex_{ICE} = \sum_{i=1}^{j} (Ex_{ICE} + M_{ICE} + F_{ICE})_i \tag{7}$$

3.1. Cost of the Externalities of the Vehicle

The externalities that have been presented will offer different values depending on whether an ICE vehicle or a V2G is used. In the first case, as represented in (8), these externalities will be closely linked to the cost of the health impact caused by ICE-based vehicles (h_{ICE}), the distance run with the vehicle (*D*), as well as the average consumption of gas ($Av_{consICE}$), carbon emissions (depicted as C_{ICE} for the ICE vehicle) and the social cost of carbon (*SCC*) during a certain period of time. These externalities are also reflected for the V2G solution in (9), where the equivalent data has been included. Health impact (h_{V2G}) and carbon emissions (C_{V2G}) are harder to measure in the case of V2G, as they are related to the energy mix from which electricity is coming and, more specifically, the amount of renewable energies present in this energy mix. The cost of the electricity used to move the vehicle (E_{cons}) can be determined by the trading operations that can be done by the owner of a V2G automobile (even though it will be usually lower that the cost of oil-based fuels). Overall, the cost of these externalities for society has been represented in a manner resembling the one used in [9], as there were concepts such as SCC or distance that had to be taken into account in the same way as it was done in this related work.

$$Ex_{ICE} = \sum_{i=1}^{j} (h_{ICE} \times D + C_{ICE} \times SCC)_i \tag{8}$$

$$Ex_{V2G} = \sum_{i=1}^{j} (h_{v2g} \times D + C_{V2G} \times SCC)_i \tag{9}$$

3.2. Cost of Yearly Fuel Consumption

There are several aspects that must be taken into account when including yearly fuel consumption in the mathematical model. For instance, the cost of the energy bought and the price set to sell it back to the market, so that the end user will use arbitrage to their advantage. This feature will be dependent on, among other aspects, two main factors: a) buying and selling actions that take place during energy cost peak or valley hours (while overall an average price for electricity may differ during the day depending on the user tariff, the V2G infrastructure will take advantage of a peak/valley hours setting, as depicted in [35]), and b) the possibility for the end user of the V2G to buy and sell energy at a suitable time for their interests. The latter implies that due to the usage of the vehicle or their end users' working schedule, they may not be able to charge completely their V2G during valley hours and sell all the electricity during peak hours. All these factors have been taken into account in the mathematical model presented in this manuscript: while energy is bought and sold in the average prices set for valley hours (Avc_{buy}) and peak hours (Avc_{sell}), real price of electricity when both purchased and sold is obtained as the combination of Avc_{buy},Avc_{sell} and the addition of four different factors that range from 0 to 1, which effectively describes the percentage of the energy that can be bought and sold during each of the two possible time periods (valley or peak hours). They are used to represent the fact that it will be very difficult for regular end users to buy and sell electricity during all-optimal time periods, so there will be just a majority of power bought and sold when it is best for the end user. They are called f_{bb} (for *factor* of energy *bought* during optimal *buying* period),f_{bs} (for *factor* of energy *bought* during optimal *selling* period),f_{sb} (for *factor* of energy *sold* during optimal *buying* period), and f_{ss} (for *factor* of energy *sold* during optimal *selling* period). The efficiency to buy and sell electricity at the suitable moment has been estimated at 80% (hence, the 0.8 value of f_{bb} and f_{ss}), so some power will have to be transferred when it is least optimal for them (estimated at 20%, hence the 0.2 value of f_{bs} and f_{sb}). This has been done so because there are some examples in literature that show how a portion of the EV charge is done in suboptimal periods of time. For example, it is shown [36] that there is some charging done halfway through the day, which is usually the daily time period when electricity prices gone from valley to peak in two-levelled tariffs. Equally, it is shown in [37] how charge estimations done can take place around 6 p.m., a time of the day that is often part of peak hours. These principles have been included in Equations (10) and (11), which represent the final cost of buying (C_{rpbuy}) and selling energy (C_{rpsell}) when suboptimal intervals are included. These equations are defined like this because it is assumed that there are basically two levels of prices with small fluctuations inside them (as seen in [32]).

$$C_{rpbuy} = Avc_{buy} \times f_{bb} + C_{sell} \times f_{bs} \tag{10}$$

$$C_{rpbuy} = Avc_{buy} \times f_{bb} + C_{sell} \times f_{bs} \tag{11}$$

Fuel costs are modelled differently depending on the vehicle that is used as a private transport mean. If the ICE-based solution is used, gas costs will be as shown in (12). It is basically the same way that diesel fuel costs are described in [9] (average fuel consumption Av_{cons} and gas price Cf_{ICE} have been used as variables), with the exception that figures used in this case correspond to private automobiles. The equation in (13) shows how yearly costs would be for the V2G solution. Unlike an ICE automobile, it relies heavily on the trading activities that are done with the energy stored in the battery of the V2G vehicle, which imply buying and selling energy (represented in the formula as E_{buy} and E_{sell}) to different costs: one to buy it—C_{rpbuy}—and a different one to sell it—C_{rpsell}. Buying prices are expected to be lower than selling ones; otherwise, the opportunity to make up for some of the expenditures will be lost). As it can be inferred, if during a certain period of time there is more energy sold than the one consumed, electricity cost will result negative for the V2G, which means that the owner of the vehicle will be obtaining a profit from trading with the electricity, rather than just reducing its costs via V2G usage. Both equations have included the inflation rates for ICE fuel and electricity (*Inff*). Lastly, since according to [17] there will be 95% efficiency when charging a vehicle via plug-in charging mode, an efficiency factor (e_f) has been introduced to reflect the small loss of charge when energy is transferred in and out of the electric vehicle.

$$F_{ICE} = \sum_{i=1}^{j} (1 + Inff)^i \times (Av_{consICE} \times Cf_{ICE}) \tag{12}$$

$$F_{V2G} = \sum_{i=1}^{j} (1 + Inff)^i \times \left(E_{buy} \times C_{rpbuy} \times e_f - E_{sell} \times C_{rpsell} \times e_f\right) \tag{13}$$

The cornerstone of the vehicle-to-grid technology is the capability to sell electricity to the power grid where it is installed, since it offers a unique selling point that cannot be found in other regular vehicles. Thus, the energy that can be sold back to the system has been accounted in (14). If a yearly period is considered regarding the energy that can be sold (E_{sell}), then the overall available energy to trade—that is to say, energy that can be sold during peak hours, as opposed to the most advisable time to purchase it, which would be valley hours—will be the remaining energy after considering two variables from all the energy that has been bought for charging the battery (E_{buy}): a) the energy consumed to move the automobile (E_{cons}) and b) the passive discharge of the battery when it is idle (*pdis*). Yearly amount of energy sold and bought from the power grid can be considered after learning past patterns in energy pricing and consumption. Information for a long-time span can be found from the transport system operator if required [38].

$$E_{sell} = E_{buy} - E_{cons} - pdis \tag{14}$$

In order to understand the previous equation, E_{buy} and E_{cons} must be defined too. The energy that is bought for the battery of the V2G will result from calculating the amount of power (Pw) purchased during a certain period of time (t). However, the degradation of the battery will take its toll during the battery lifetime, resulting in declining energy storage capabilities. In addition, the passive discharge of the battery must also be born in mind. While it is negligible in the short term, its effects are more noticeable during the whole lifespan of the battery. Lastly, the difference between the nominal and the actual battery charge values must also be considered. These two latter variables are hard to quantify and no work from the literature seems to portray them in an accurate manner in mathematical models for V2G technology. As far as the V2G model is concerned, they have been included as Dg (degradation factor for the battery). When numerical values are used to evaluate the model, the maximum discharge speed of the battery will also have to be considered as a non-functional requirement, as no battery can provide an immediate amount of limitless energy. Due to this, Dg will have a role in the model, even though differences may not be that significant according to D. Wang et al. [39] or H. Ribberink et al. [40]. Battery degradation for purchased energy has been included in (15).

$$E_{buy} = \sum_{i=1}^{j} (Pw \times t)_i \times (1 - Dg) \tag{15}$$

Battery degradation has been estimated by the authors of this manuscript to be at 1.25% of its total capacity per year so it can be included with more accuracy in the mathematical model. The reviewed literature shows extreme disparity regarding this value, with some sources claiming that it will degrade up to 10% after 160,000 miles for an electric vehicle [41]. However, battery degradation considered for this scenario has been regarded as significantly higher, as a) suboptimal charge and discharge behavior patterns from the vehicle owners must be taken into account, and b) V2G usage of an electric vehicle implies a heavier utilization of the vehicle battery. A more realistic approach is found in [42], where a thorough V2G-based experiment was run with experimental lithium batteries showing that they would reach their end of life (EOL), regarded to be the point when the battery has lost 20% of its original maximum capacity retention, after 3000 cycles of charge and discharge. For the purpose of this mathematical model, it has been estimated that, on average, 1000 cycles will take place every year for the V2G solution (as described in [43]), and after eight years the battery total capacity will be depleted a 20% and have to be replaced with a new one. Thus, battery degradation is defined as represented in (16).

$$Dg = 0.0667 \times \sum_{i=1}^{j} i \tag{16}$$

The energy that is consumed by the V2G solution can be described as the average energy consumption of the vehicle during a specific distance (E_{cons}). As explained before, passive energy losses have been included as the *pdis* parameter.

3.3. Cost of Maintenance

Although it is not bound to happen inevitably, the battery used in the EV-V2G is very likely to eventually have to be replaced. However, it does not necessarily mean that the vehicle owner will pay for the full replacement if the vehicle has been acquired under a battery leasing agreement. Therefore, there are two possible options: if the vehicle and the battery are purchased, battery replacement costs will have to be considered; with the technology available today in commercial products, it is unlikely that a vehicle battery will outlive the vehicle itself. The other option, though, is that the vehicle manufacturer leases the batteries to the vehicle owner during a certain time period. In this way, battery reposition could be regarded as a periodic payment ($Bleas_i$) done during the lifetime of the vehicle. This latter scenario is modelled in (17) as $Bleas_i$; while this is not the default choice for consumers buying an electric vehicle, it is a feature usually overlooked in other models for V2G, so it has been included in this analysis. When price data are used to estimate the cost differences between acquiring and leasing the battery in the V2G, $Bleas_{tot}$ would be used as the maintenance cost for rented batteries, whereas $Batr$, added in (19), will be used as the parameter representing the cost of a battery replacement when the battery is purchased with the EV.

$$Bleas_{tot} = \sum_{i=1}^{j} Bleas_i \tag{17}$$

Lastly, maintenance costs have been included in the model as a way to evaluate the differences between the two kinds of vehicles. The ICE vehicle (18) makes use of a maintenance rate ($Drate_{ICE}$), in a way that resembles the one presented in [9], but using private transport rather than a school bus. Labor costs of refilling the fuel (Lab) and distance (D) have also been included. Furthermore, the equation conceived for the V2G solution (19) is making use of an equivalent rate ($Drate_{V2G}$) and a distance D and the cost of one battery replacement ($Batr$). Taking into account the average lifetime of EV vehicles and of their batteries before a replacement (which can be estimated at roughly eight years according to the period warranty used in most car manufacturers [43,44]), it is more likely that a new vehicle will be acquired rather than a new whole battery is bought more than once. The equation that has been added as (19) can be modified to consider how battery costs impact the maintenance of a V2G solution when the battery is leased instead of purchased (20). Note that both kinds of vehicles will require the payment of yearly insurance costs (which has been represented by Ins). However, according to [45], their payment can be regarded to be the same for them.

$$M_{ICE} = \sum_{i=1}^{j} (Drate_{ICE} \times D + Lab + Ins)_i \tag{18}$$

$$M_{V2G} = \sum_{i=1}^{j} (Drate_{V2G} \times D + Ins)_i + Batr \tag{19}$$

$$M_{V2G} = \sum_{i=1}^{j} (Drate_{V2G} \times D + Ins + Bleas)_i \tag{20}$$

4. Numerical Assessment

The equations of the mathematical model described previously have been put to use for three different use cases, namely, professional drivers (that is to say, people that drive as a way to make their living), frequent drivers (people that drive on a usual basis), and occasional drivers (people that drive rarely), under certain considerations and assumptions as described in the following subsections. Most references and subsidy figures that have been used are relative to the United States of America, due to the fact that it is one of the places where the amount of information was plentiful enough to obtain the data used in this study. Specifically, data for professional drivers was very reliable as it was based on statistics from taxi drivers that are offered online freely. The definition of these use cases is pivotal for the study that has been carried out, as the usage that is done of the V2G solution differs greatly in each of them. Depending on the usability of the vehicle for travelling, V2G capabilities will become prominent. For example, the greater amount of distance that a V2G solution works, the lower energy will be left to trade it when it is suitable.

4.1. Considerations

Table 3 contains the information regarding how the variables that have been introduced in the previously detailed equations have been given numeric values according to the existing related work. Some of those variables do not change in the three scenarios but many other do so, as they are closely linked to the case study involving the vehicle (fuel, distance driven, etc.). In this manuscript, the price of a V2G solution has been estimated to be $7500 higher than an ICE-powered counterpart; as far as the United States are concerned, financial aid of up to that quantity is offered to the buyers of a full EV solution in some regions [3,46], so it has been included as an EV overprice in the model.

As for the battery replacement, it has been regarded as an average value of the figures found in [47] and [48]. The result has been depicted in Table 2, which considers four vehicle models. Several car models are considered in this chart, according to the information provided in [47]. It has been considered that the data in [47] can be divided into a best case scenario with a 40 kWh battery, where the Nissan Leaf owner does not require to pay any extra other than the battery replacement, and a worst case scenario where the Nissan Leaf owner must pay both for the special adapter kit ($225) and labor costs of $1000 when the old battery is exchanged with the new one (also with a 40 kWh battery). These results demonstrate alignment with other studies that show battery cost to have been declining during the last decades, such as the one shown in [49].

Table 2. Average battery cost per kilowatt/hour.

Vehicle	Battery Cost ($) per Kilowatt/Hour	Reference
Nissan Leaf best case scenario	$5499/40 = 137.45 $/kWh	[47]
Nissan Leaf worst case scenario	($5499 + $1000 + 225)/40=168.1 $/kWh	[47]
Chevrolet Bolt EV	205 $/kWh	[48]
Tesla Model 3	190 $/kWh	[48]
Average	**175.14 $/kWh**	N/A

Considering that a vehicle battery of 40 kWh has been used for this manuscript, the cost of its replacement used in the numerical assessment results in **175.14 $/kWh × 40 kWh = $7005.6**.

It must be noted that the figures corresponding to professional, frequent, and occasional drivers are strongly related to the information that has been inferred from several sources present in this manuscript, such as [50] and [58]. It is said in [50] that taxi cabs can be driven up to 70,000 miles, whereas it is claimed in [58] that average miles travelled by a vehicle are 11,370. This is the mileage that has been defined for frequent drivers (people who drive a car often enough to require it during a significant amount of days of the year but do not make a living out of using automobiles). In order to strengthen the criteria used to have an accurate view of the mileage that defines each case study (professional, frequent, and occasional drivers) two more references have been studied. On the one hand, it is said in [66] that 2813 gallons per car and per year are consumed by taxi drivers, who represent the archetypical professional driver use case. On the other hand, it is claimed in [67] that 524 gas gallons are used yearly per vehicle. Despite these figures are prone to change as time goes by or depending on boom or bust economic cycles, they can be used as representative values of mileage and gas consumption. Consequently, and considering the ratio of gas usage existing between professional and frequent drivers (2813/524 = 5.368) it has estimated that a) since frequent drivers drive 11,370 miles per year and b) mileage figures for professional drivers are unlikely to go beyond 70,000 miles per year, professional driver mileage can be estimated as 11,370 × 5.368 = 61,033 miles per year. As it will be described in use case C, due to the data presented in [68], it has been estimated that occasional drivers make use of automobiles a quarter of time (which has been correlated to mileage) than frequent drivers. Another aspect to consider is the relationship between the mileage for each use case and the energy being used in every one of them. It has been estimated that, according to the figures that can be obtained from [50] and [58] and the ratio of gas usage explained in the previous paragraph, yearly mileage will be of 61,033 miles for a professional driver, 11,370 miles for a frequent driver and 2842.5 miles for an occasional driver. Additionally, if the average figures that can be extracted from [61] are considered as well, battery consumption would be of 20,301.67 kWh for professional drivers, 3781 kWh for frequent ones and 945 kWh for occasional ones. Furthermore, there is a certain battery degradation coming from using the V2G functionalities of the enhanced EV which is far more significant than usual wear off in an EV battery. Consequently, the energy that can be traded every year depends on (a) the amount of energy available for trade (the more frequent a person drives, the higher amount of energy is used for driving; hence, V2G energy costs will be overall higher as lower profits can be made from trading) and (b) battery degradation (as time goes by, capacity of the battery will shrink). These considerations are especially important for Tables A5–A7, where profitability of the solution is described in relation to whether battery degradation is present or not.

4.2. Case Study A: Professional Drivers

This case study involves people whose main job implies driving or taking passengers in a private-like means of transport (taxi drivers are the most typical example). This kind of job implies that there will be high costs in consumed fuel and maintenance for ICE-based vehicles. As represented in Table 3 and mentioned earlier, the costs and usage for professional drivers have been calculated considering those according to [53], namely, the yearly average consumption of gas is 2813 [63]/524 [67] = 5.368 times the one made by the frequent drivers even if, as mentioned before, there are cases where taxi cabs are driven up to 70,000 miles per year [45]. The following figures, adjusted to inflation, have been obtained:

$$Ex_{ICE} = \$68,843.39 \quad Ma_{ICE} = \$404,203.13$$

$$F_{ICE} = \$90,926.12$$

It can be inferred that the total costs for a professional driver using an ICE automobile for twelve years are the following ones:

$$C_{totICE} = \$35,285 + \$563,972.65 = \$599,257.65$$

If a V2G solution is used instead of an ICE-based vehicle, the results obtained when adjusted to inflation are different and overall lower:

$$Ex_{V2G} = \$12,278.20$$

$$Ma_{V2G} = \$90,887.29 \text{ total (with a 40 kWh battery)}$$

$$Fc_{V2G} = \$19,602.71$$

$$V2G \ conversion + Cost \ of \ the \ installation = \$1936$$

From these figures, it is calculated that the total costs for a professional driver using a V2G automobile are:

$$C_{totV2G} = \$42,785 + \$1936 + \$7005.60 + \$12,278.20 + \$19,602.71 + \$90,887.29 = \$174,494.79$$

Table 3. Variables included in the mathematical model.

Variable	Description	Value (ICE)	Value (V2G)
$Av_{consICE}$	Yearly average consumption of gas to move the ICE vehicle	2813 */524 **/131 *** gallons [50]	–
Avc_{buy}	Average cost of bought energy	–	9.35 cents/kWh (off-peak hours) [51]
Avc_{sell}	Average cost of sold energy	–	15 cents/kWh (peak hours) [51]
$Batr$	Battery replacement	–	$7005.60 [47,48], Table 2
$Bleas$	Yearly battery leasing	–	ca. $140 × 12 [52]
$Bleas_{tot}$	Total cost of battery leasing	–	$Bleas$ × 12
$Capex_{ICE}$	Cost of acquiring an Internal Combustion Engine-powered vehicle	$35,285 [53]	–
$Capex_{V2G}$	Cost of acquiring an Vehicle-to-Grid-powered vehicle	–	(2)
C_{ICE}	Yearly carbon dioxide emissions for an Internal Combustion Engine vehicle	19.6 (8.89 kg) lbs/gallon [54]	–
C_{V2G}	Yearly carbon dioxide emissions for a Vehicle-to-Grid-powered vehicle	–	149.25 [55] × 40 = 5,97 MT/12 years = 497 kg/year
Cf_{ICE}	Cost of the fuel for an Internal Combustion Engine-powered vehicle	$2.176/gallon [56]	–
C_{EV}	Cost of acquiring the Electric Vehicle	–	$42,785 [53,57]
C_{rpbuy}	Real price of bought energy	–	(12)
C_{rpsell}	Real price of sold energy	–	(13)
C_{totICE}	Total costs of purchase and usage of the Internal Combustion Engine vehicle	(1)	
C_{totV2G}	Total costs of purchase and usage of the Vehicle-to-Grid automobile		(2)
D	Yearly distance	61,033 */11,370 **/ 2842.5 *** miles [58]	61,033 */11,370 **/ 2842.5 *** miles [58]
Dg	Degradation factor of the battery (State of Health)	–	(16)
$Drate_{ICE}$	Average maintenance rate per mile by an Internal Combustion Engine-powered vehicle (medium sedan)	$0.5762−$0.116 [59] = $0.4602	–
$Drate_{V2G}$	Average maintenance rate per mile by a Vehicle-to-Grid-powered vehicle	–	$0.09204 (1/5 of [9,59])
E_{buy}	Yearly amount of energy bought	–	2304 kWh × 1000 battery cycles (from [60])
E_{cons}	Yearly Amount of energy consumed	–	20,301.67 */3781 **/ 945 ***kWh [61]

Table 3. *Cont.*

Variable	Description	Value (ICE)	Value (V2G)
e_f	Efficiency factor for charging a vehicle via power cable	–	0.95 [17]
E_{sell}	Yearly amount of energy sold	–	(13)
EX_{ICE}	Externalities for an Internal Combustion Engine-powered vehicle	(8)	–
EX_{V2G}	Externalities for a Vehicle-to-Grid (V2G)-powered vehicle	–	(9)
F_{ICE}	Gas costs for an Internal Combustion Engine-powered vehicle	(12)	–
F_{V2G}	Energy costs for a Vehicle-to-Grid vehicle	–	(13)
f_{bb}	Factor for energy purchase in optimal buying hours	–	0.8 *****
f_{bs}	Factor for energy purchase in optimal selling hours	–	0.2 *****
f_{sb}	Factor for energy sell in buying optimal hours	–	0.8 *****
f_{ss}	Factor for energy sell in selling optimal hours	–	0.2 *****
h_{ICE}	Per-mile cost of the health impact caused by the electricity consumed by the ICE vehicle	$0.07 (estimated from [9,10])	–
h_{V2G}	Per-Mile cost of the health impact caused by the electricity consumed by the V2G vehicle	–	$0.0149 [9]
Heq	Cost of the installation of the required equipment to charge the Electric Vehicle	–	$1200 [61]
Inf	Average inflation 2008−2019 (US)	1.76% [62]	1.76% [62]
$Inff$	Inflation rate on fuel	3.8% [57]	1.9% [9]
Ins	Yearly cost of insurance	$1251 [58]	$1251 [58]
Lab	Yearly fuel refill labor	$1207.80 */225 **/56.25 *** [9]	–
M_{ICE}	Maintenance costs of an Internal Combustion Engine-powered vehicle	(18)	–
M_{V2G}	Maintenance costs of a Vehicle-to-Grid vehicle	–	(19)
$Opex_{ICE}$	Operational costs to have the ICE vehicle in fully working condition	(4)	–
$Opex_{V2G}$	Operational costs to have the V2G vehicle in fully working condition	–	(5)
$pdis$	Passive discharge of the battery	–	5.59%/30% **/ 120% **** [63]
Pw	Amount of power purchased	–	$Ebuy/t$
SCC	Social Cost of Carbon	$37.20/MTCO2e ([64], calculated for 2016 dollars)	$37.20/MTCO2e ([60], for 2016 dollars)
t	Period of time	–	Variable; 12 years for Section 5
$Vcons$	Cost of conversion to V2G technology	–	$736 [65]

* Professional drivers, ** Frequent drivers, *** Occasional drivers **** 120% represents that a charge cycle and a fifth of another one are lost ***** Chosen as a plausible hypothesis.

As it can be inferred from the previous calculations, it can be seen that the V2G solution is far more economically efficient for a professional driver in the long term than an ICE vehicle. The graphical representation of the cumulative result that has been calculated for each of the years is displayed in Figure 2. At the same time, Table A1 is showing in the Appendix A how numerical calculations vary on a yearly basis as well.

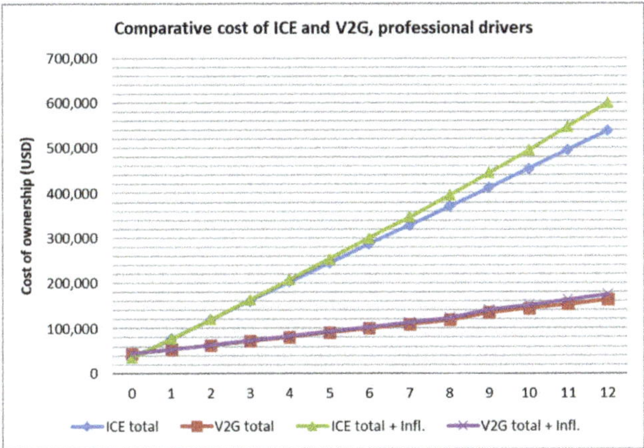

Figure 2. Graphical representation of the calculation results for professional drivers.

If the results that have been obtained are separated so that operational costs can be considered more accurately, it can be seen how despite a) a higher Capex if an EV is purchased; b) the required infrastructure to make the EV work as a V2G solution; and c) a battery renewal, the operational costs of the ICE vehicle are far higher in this scenario starting from year 1, mostly but not only, due to the maintenance costs required to have the ICE working satisfactory. This is the key advantage that the V2G solution has, which makes it economically far more advisable under these circumstances if compared to the ICE alternative. The graphical comparison of Opex costs has been displayed in Figure 3. Note that the battery replacement has been included as an Opex-related expenditure, so it is present in the cumulative figures. Moreover, it is considered that the first year of usage (year 0 in the previous graphs) there are not operational costs, which start being added at year 1. That is why this and the other equivalent graphs show year 0, whereas Opex-related ones do not.

Note that the previous results have been obtained under conditions deemed as "suboptimal" in terms of cost of energy purchase and sell. That is, all the energy has been bought during valley hours and sold during peak hours in a proportion of 80/20. This implies that according to the parameters that have been included in (8), (9), (10), and (11), it has been considered that 80% of the energy purchased was done so during valley hours and the other their during peak hours (so that $f_{bb} = 0.8$ and $f_{sb} = 0.2$), whereas 80% of the energy was sold during peak hours and the other 20% during valley ones (and thus, $f_{ss} = 0.8$ and $f_{bs} = 0.2$). Furthermore, small losses when charging and discharging the vehicle may result in a loss of electricity during these procedures (hence, $e_f = 0.95$, as described in Table 3. Lastly, the degradation of the battery has also been considered when doing the calculations according to the mathematical model. Hence, the energy that has been estimated to be sold every year decreases over time in the rate established in (14).

Overall, the proportion that is sold during each of the time periods will depend on the available power to operate in the market and the availability of the user of the V2G solution. There are two important aspects that can be inferred by all these calculations: under a time period of longer length than the one used here (12 years) the advantages of the V2G solution over the ICE one will be even more notorious as the ones portrayed in this time span, as cost differences between both of them are always unfavorable for the ICE vehicle. Additionally, a suboptimal scenario has little to no influence in the calculations done for both battery rental and acquisition, as the former one will become unfavorable in the long term, according to the figures obtained for battery rental that have been introduced in Section 4.5.

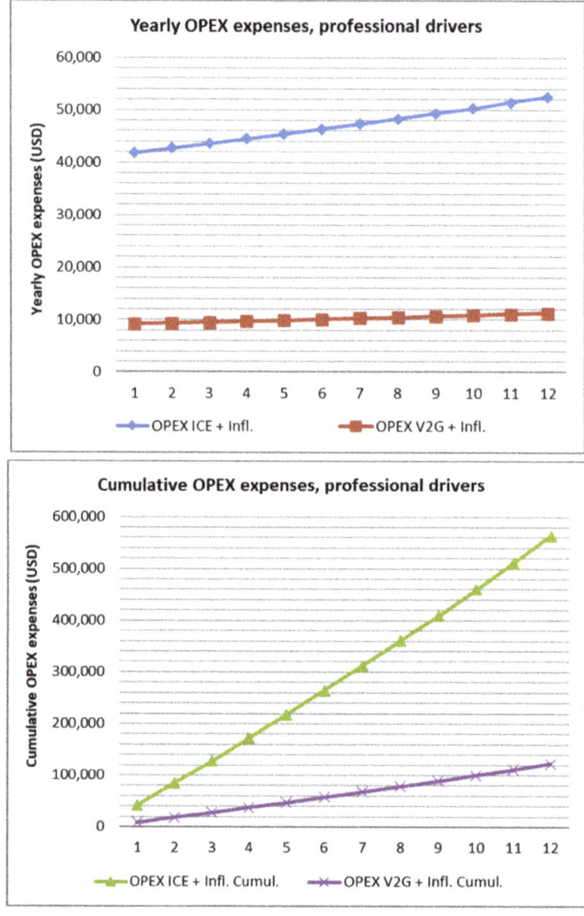

Figure 3. Yearly and cumulative OPEX expenses between an ICE vehicle and a V2G, professional drivers.

4.3. Case Study B: Frequent Drivers

The most representative situation that can be conceived for this use case is a freelance worker with a specific job that make them travel a significant distance every day (self-employed positions, etc.), but do not use driving as their business core. Frequent drivers are regarded in this numerical assessment as the average group of people, so they have been assigned the default figures that have been found in literature.

If the same calculations that were done previously are repeated for this use case, the next results are obtained adjusted to inflation:

$$Ex_{ICE} = \$12,824.83 \ Ma_{ICE} = \$88,770.04$$

$$F_{ICE} = \$16,937.54$$

Thus, the following costs will have to be assumed by the owners of an ICE vehicle during its lifetime will be

$$C_{totICE} = \$35,285 + \$118,532.40 = \$153,817.40$$

Should a V2G solution be used, the results would be

$$Ex_{V2G} = \$2486.41$$

$$Ma_{V2G} = \$30,401.65 \text{ (with a 40 kWh battery)}$$

$$Fc_{V2G} = \$ -9233.58$$

$$V2G\ conversion\ +\ Cost\ of\ the\ installation\ =\ \$1936$$

Note that the fuel (electricity) cost for the V2G vehicle is negative for this case study, due to the fact that selling the energy surplus is creating a profit for the end users of the vehicle, to the point that energy trading results economically advantageous for the end user in terms of energy costs. This is due to the fact that energy is being bought and sold in a proportion that makes the sold energy more economically significant in absolute values than the one that is being bought. Therefore, the engagement in energy trading for clients using the V2G solution becomes profitable, as the usage of vehicle-to-grid technology makes possible decreasing the costs of using an electric vehicle when the unused power is sold back. Thus, the total costs for a frequent driver that owns a V2G automobile would be:

$$C_{totV2G} = \$42,785 + \$1936 + \$7005.60 + \$30,401.65 - \$9233.58 + \$2486.41 = \$75,381.09$$

Figure 4 depicts the graphical representation of the obtained results whereas the calculations that have been carried out are presented in Table A2 of the Appendix A. Note that Figures containing graphs show a sudden non-linearity for the costs of the V2G solution event between years 8 and 9. This is due to the fact that it has been estimated that it will be the moment when battery from the V2G solution will have to be eventually replaced, so expenses rise accordingly to the $7005.60 that have to be spent. Moreover, benefits towards the V2G solution do not start right away but after year 2, thus showing that this scenario is less advantageous due to the lower costs for the ICE vehicle.

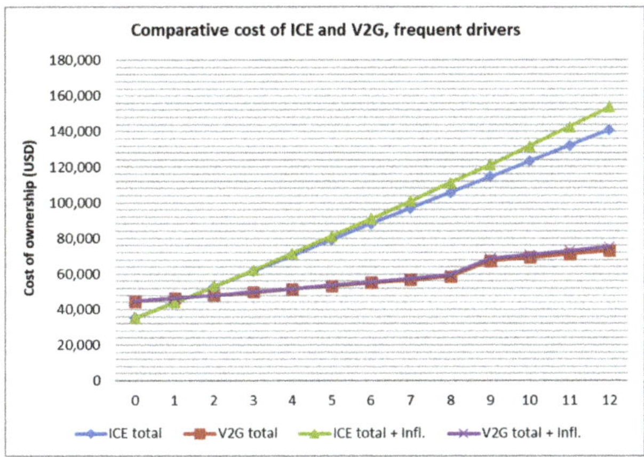

Figure 4. Graphical representation of the calculation results for frequent drivers.

As mentioned previously, Figure 5 depicts the differences in operational costs between the V2G and the ICE vehicle. Albeit with a smaller gap resulting from the lesser usage of the automobiles, the results are essentially replicated for this case study: yearly expenses, and cumulative ones when the battery replacement costs are included, are lower for the V2G than the ICE vehicle.

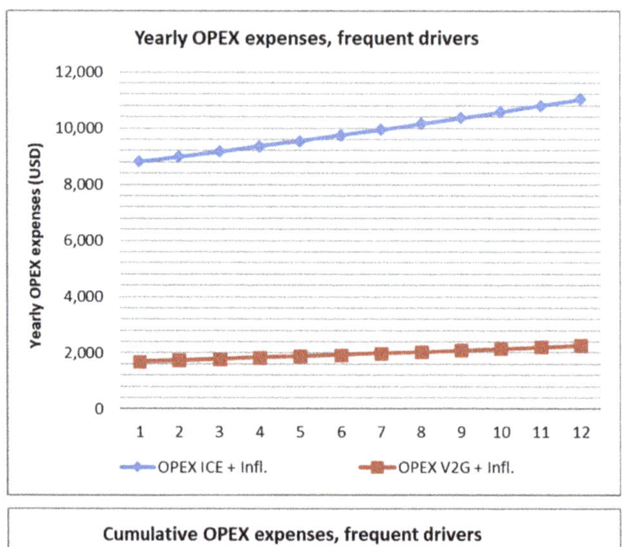

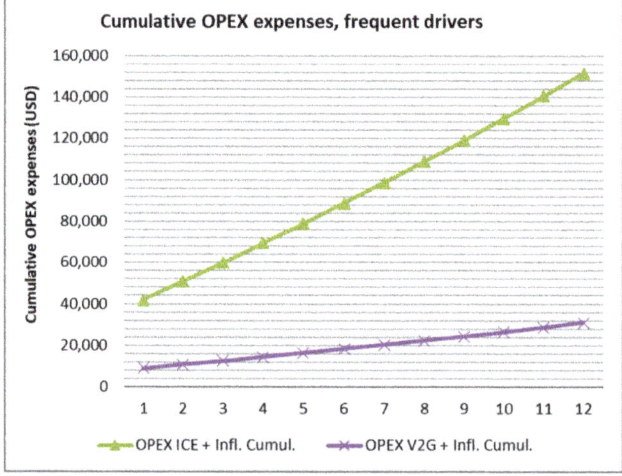

Figure 5. Comparison between yearly and cumulative OPEX expenses between an ICE vehicle and a V2G, frequent drivers.

As stated previously, purchasing and ICE vehicle results in a worse economy cost for the prosumer if compared to acquiring a full V2G solution (as the former implies higher costs of fuel, maintenance and externalities during the vehicle lifetime). Fuel consumption falls considerably for the V2G in this scenario, as more energy is used for trading operations. The suboptimal scenario is also used for this use case with the same set of variable values that was employed before (f_{bb} =0.8, f_{sb} =0.2, f_{ss} =0.8, f_{bs} =0.2, and $e_f = 0.95$). As in the previous case, despite obtaining a worse result with a suboptimal scenario where energy is neither bought nor sold under the best possible circumstances, it is still better than the one that would be obtained with the ICE solution.

4.4. Case Study C: Occasional Drivers

An occasional driver has been defined with the same criteria that was done in [68] and [69], that is to say, "A driver who operates a vehicle less than 25 percent of the total miles put on the car during a year". Consequently, it can be assumed that, when compared to frequent drivers, an occasional driver will use the vehicle one fourth of the time a frequent driver would, so all the expenses have been considered to be one fourth of the ones calculated in the previous case study. As far as the ICE automobile is concerned, results adjusted to inflation are as follows:

$$Ex_{ICE} = \$3206.21 \quad Ma_{ICE} = \$34,607.92$$

$$F_{ICE} = \$4234.38$$

Therefore, the resulting budget for an ICE vehicle owned by an infrequent driver would be

$$C_{totICE} = \$35,285 + \$42.048,52 = \$77,333.52$$

Thus, operational costs have become lower than the purchase of the vehicle itself. If a V2G solution is used, results obtained are

$$Ex_{V2G} = \$805.70$$

$$Ma_{V2G} = \$20,015.83 \ (\text{with a 40kWh battery})$$

$$F_{V2G} = -\$14,128.02$$

$$V2G \ conversion + Cost \ of \ the \ installation = \$1936$$

As it happened before, the fuel costs for electricity in this case are negative. What is more, since there is more electricity available to be sold (as it is used to a lesser extent by the vehicle), saving costs become even more prominent than in the previous case study. The final costs would be as follows:

$$C_{totV2G} = \$42,785 + \$1936 + \$7005.60 + \$805.7 + \$20,015.83 - \$14,128.02 = \$58,419.48$$

As it was done in the previous cases, the suboptimal scenario has been used with the same set of variables ($f_{bb} = 0.8$, $f_{sb} = 0.2$, $f_{ss} = 0.8$, $f_{bs} = 0.2$, and $e_f = 0.95$) energy costs are higher than in the optimal scenario. Unlike previous case studies, the EV-V2G is not as in clear advantage over the ICE vehicle in terms of expenses as it was before. What is more, it would not be until the fourth year of ownership that the V2G solution shows a better performance when compared to the ICE automobile. The main reason for this is that, although the V2G solution decreases its costs as long as the battery is kept the same, as soon as the latter is replaced, costs rise above the ICE level, thus closing the gap between the two kinds of vehicles. The graphical representation of this fact is shown in Figure 6.

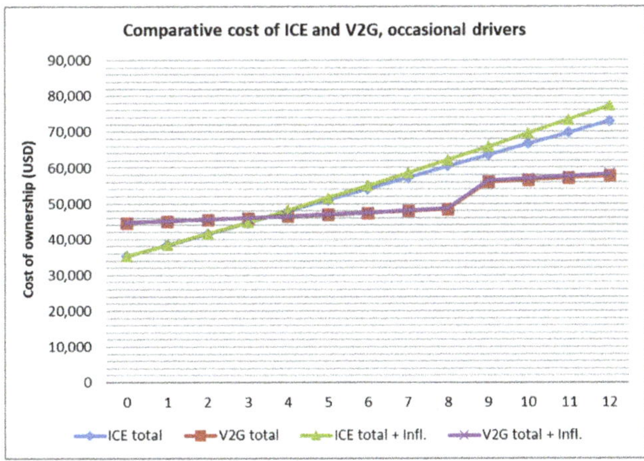

Figure 6. Graphical representation of the calculation results for occasional drivers.

Additionally, Figure 7 shows a comparison between operational costs between the V2G and the ICE options for occasional drivers. As in previous cases, yearly and cumulative expenses for operational costs are lower when the EVV2G is used instead of the ICE. However, the differences are less significant this time, to the point that the higher purchase cost of the EVV2G and its frequent battery replacement make it harder to justify using it. Interestingly enough, if the battery replacement is not taken into account, OPEX for the V2G shows almost stagnant figures. This is due to the fact that the V2G is used so little that it is highly available to trade energy in favorable terms with the overall grid system, and it results in a profit for the end user who owns it.

Table 4 shows a numerical summary of the cost of these three use cases.

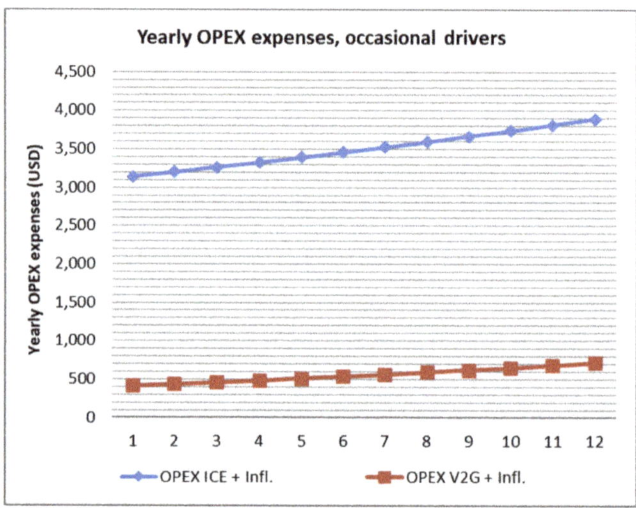

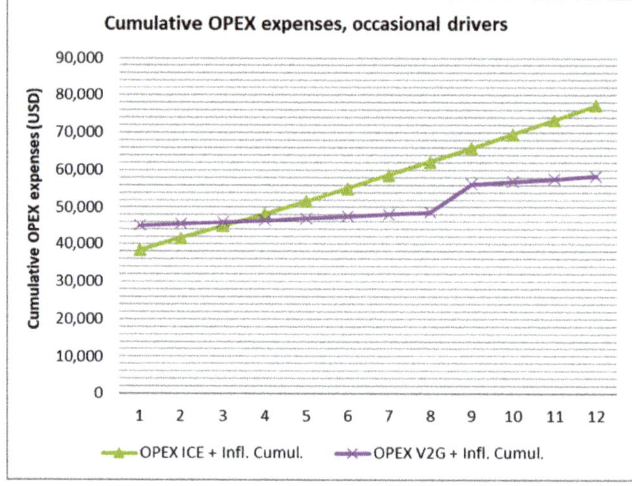

Figure 7. Yearly and cumulative OPEX expenses between an ICE vehicle and a V2G, frequent rivers.

Table 4. Costs summary of ICE and V2G.

		Professional Drivers	**Frequent Drivers**	**Occasional Drivers**
Externalities	ICE	$68,843.39	$12,824.83	$3206.21
	V2G	$12,278.20	$2486.41	$805.70
Maintenance	ICE	$404,203.13	$88,770.04	$34,607.92
	V2G	$90,887.29	$30,401.65	$20,015.83
Fuel	ICE	$90,926.412	$16,937.54	$4234.38
	V2G	$19,602.71	−$9233.58	−$14,128.02
CAPEX	ICE	$35,285	$35,285	$35,285
	V2G	$44,721	$44,721	$44,721
Total	ICE	$599,257.65	$153,817.40	$77,333.52
	V2G	$174,494.79	$75,381.09	$58,419.48

4.5. Comparison between Battery Rental and Battery Ownership

If it is chosen to purchase an EV where the battery is rented rather than acquired with the same vehicle, the average costs obtained after twelve years (adjusted to inflation) with one battery replacement according to the mathematical model are as follows:

$$CAPEX_{V2G} = \$51,726.60 \text{ (with a 40 kWh battery purchase)}$$

$$CAPEX_{V2G} = \$56,586.66 \text{ (with 40 kWh battery rental)}$$

The yearly comparison of each option has been depicted in Figure 8, whereas the calculations themselves have been placed in the Appendix A, Table A4.

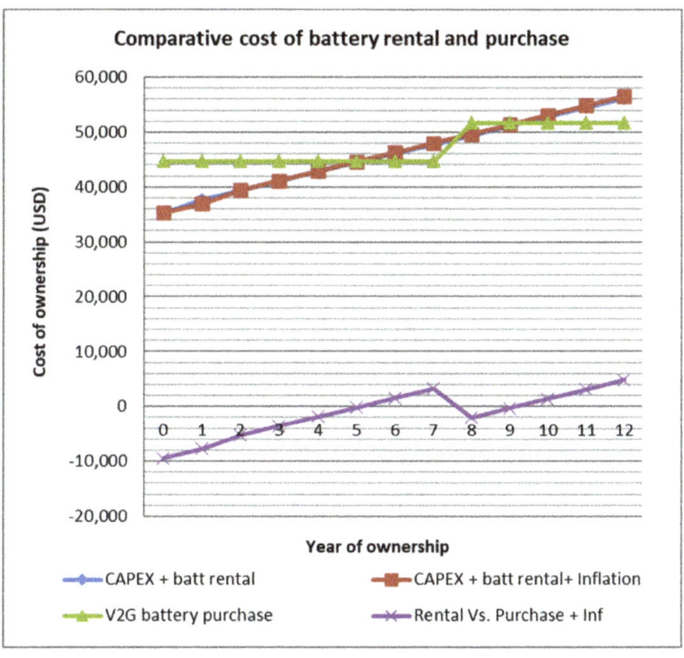

Figure 8. Graphical representation of battery rental costs versus battery purchase ones.

If these results are compared thoroughly, it can be seen that the battery purchase option becomes more advisable to use in the long term when the EV has been enabled to make use of V2G technology, whereas it is the opposite for shorter term ownership (5 or less years). This is due to the fact that the battery installed in the vehicle becomes depleted at a faster rate than a conventional EV which makes no use of V2G, thus being more likely to have its battery replaced once during its usage timespan. Should the battery not require to be replaced, then battery ownership option would result more competitive. However, it must be taken into account that usually, no manufacturer that offers battery rental as an option expects its customers to use it as part of the equipment of V2G technology. Probably, manufacturers would put restrictions to their usage if end users were openly planning to use their vehicles with this kind of technology.

It must be noted that, with the battery rental option, all the considerations done previously regarding battery degradation are still valid. Battery will degrade at a similar rate regardless of how the owner pays for its usage, as the components and chemical reactions that make it work remain the same in both cases. That is why battery replacements are considered under the purchase option, as any V2G solution will make use the car battery intensively (due to the very nature of V2G, which demands more frequent energy discharges and recharges than a EV battery used one-way only) and will have to be replaced after a relatively short amount of time, whereas a rented battery will degrade with the same parameters but the cost of its replacement will not have to be assumed by the end user.

Nevertheless, the scenario where the rented battery of the EVV2G is replaced every year could be put forward as another part of this study. In this case, yearly degradation could be considered as zero (as the battery would be replaced every year) and greater amounts of energy could be purchased and sold, due to the battery capacity being maintained during the lifetime of the V2G solution. In this case, more energy would be available for trading operations, thus resulting in an increase of the profitability of the rented battery V2G solution. Yearly surplus in energy availability would progressively increase compared to a purchased battery, as shown in Table 5. Should it be assumed that only yearly degradation is taking place with the battery rental option, results would vary in favor of the latter, but the overall tendency would be the same: for long periods of time, battery purchase would be more efficient than battery rental.

Table 5. Profit difference between battery degradation and non-battery degradation. Assuming a yearly replacement of the battery, it would be the difference of battery renting vs. battery ownership.

Year	Difference Degree/No Degree	Difference Purch./Rental No Deg.
1	0.00	−9436
2	16.40	−7726.43
3	33.42	−5282.34
4	51.05	−3588.75
5	69.29	−1895.22
6	88.14	−201.75
7	107.60	1491.66
8	127.68	3185.03
9	148.36	−2127.24
10	169.66	−433.92
11	191.57	1259.37
12	214.10	2952.67

The yearly figures that would be obtained would be as portrayed in Tables A5–A7 for professional, frequent and occasional drivers. Note that regardless of the kind of driver that makes use of the solution the difference is the same in every case, as the increase in energy available is due to the same reason (the same improvement in energy used for trade operations).

5. Impact on Grid Utilities

The previously described model has been conceived for its usage in V2G solutions that become part of the entities able to provide power to sell and purchase at the electricity markets. For example, in [70] it is stated that despite the dominant trend in charging V2G is using off-peak hours, coincident user patterns can pose a threat for power system components both when charging vehicles and injecting power to the grid. The authors claim that it would be possible to overcome that problem by assessing the suitable V2G penetration level for optimal operation and precisely planning the V2G behavior on the distribution system. Furthermore, [71] describes how the addition of V2G parking lot facilities creates additional energy losses in the feeders of the electric utility owners derived from the behavior of reactive power injection and the load patterns of the users. A way to minimize this issue would be locating optimally a parking lot along the aforementioned feeder.

Additionally, V2G technology can be used to make power consumption more regular and avoid the peaks and valleys that create issues for the power grid: because of the tendency of end users to charge their vehicles in off-peak hours and not to demand that energy during the peak ones, V2G effectively becomes a way to enhance peak shaving and valley filling curves of energy demand. It is stated in [72] that combining V2G solutions with energy storage and photovoltaic electricity generation could result in a reduction of up to 37% during peak periods. Furthermore, in [73] it is claimed that by following a strategy based on comparing a forecasted load curve with another one based on forecasting available charge and discharge power peak shaving can be controllable and real, thus proving that V2G can be used as a way to create a more balanced demand of electricity. What is more, it is said in [74] that a high penetration of EVs is very likely to demand a stronger and more reliable power network; according to the authors of that manuscript, transformer replacement costs reach 72% of the total deployed transformers value with an EV penetration of 50%. However, this manuscript does not consider that the added power for that V2G can be brought to the power grid. Moreover, it is stated in [75] that distribution transformer may experience a measurable loss of life resulting of the increased strain that power demanded by plug-in hybrid electric vehicles (PHEVs) may produce. In this case, this study deals mostly with how PHEVs interact with the grid, describing the possibility of using V2G technology as part of the applications of PHEVs. The results shown in this manuscript demonstrate that V2G can be a viable solution for end users to obtain an economical benefit with their vehicles. However, they also show that the status of development in batteries makes profitability difficult, as the rapid degradation and their relative expensive cost depletes most of the benefits that could be obtained. Arguably, other options implying reducing the costs of purchasing an EV and converting it into an EVV2G could be satisfactory, but such a solution looks unlikely to happen in the short term. Overall, it is assumed that the existence of V2G solutions will strain the power grid in the short term, but there are advantageous solutions that can be integrated in the resulting smart grid. A typical solution that could come to this system advantage would be the integration of V2G technology with the other components of the power grid via middleware architectures [76] so they can be seamlessly included in such heterogeneous deployments.

6. Conclusions

According to the study done in this manuscript, purchasing an EV or V2G automobile and having the battery leased, instead of bought altogether with the vehicle, is economically inefficient for periods of time longer than five years, except for a comparatively brief period of time after the vehicle battery is replaced. On the other hand, even if charging a V2G solution during long periods of time might not be a solution for some drivers and electricity has to be purchased sometimes during peak hours or sold during off-peak slots of time, V2G technology is still more economically efficient overall when compared to ICE vehicles. This can be seen in the results that have been obtained in the study done, where it is estimated that a V2G becomes almost immediately more economically efficient for professional drivers, whereas the same happens for frequent drivers after one year. As for occasional drivers, it is estimated to take from 3 to 4 years for V2Gs to be more economically efficient than ICE-powered vehicles. Despite differences in the periods of time depending on the profile, there are two tendencies: a) the higher the usage of a vehicle, the faster it turns into a more economical choice to use V2G technology and b) the longer the time a V2G vehicle is used, the more convenient it is to buy the battery rather than renting it. Regardless of the more intense battery degradation and mandatory battery replacement that must be done in a V2G vehicle, it will result in a more economic usage in the medium-to-long term. Even if the battery had to be replaced two times in the timespan used in this study, results would still be favorable for V2G vehicles over ICE-powered ones.

That said, although V2G technology is more cost-efficient in the long term than ICE solutions, batteries are still the main bottleneck for greater profits, as they impose limits to the savings that can be done from purely maintenance costs. The potential profits (or at least, expenses reduction) that can result from applying this technology are strongly linked to battery degradation and battery costs. The first one puts a severe strain on the profitability of V2G as part of the appeal of acquiring an EV. According to the study done in this manuscript, the fact that batteries will have to be replaced in the EVV2G once limits the practical applicability of this technology to exploit it in a profitable manner. Battery costs, on the other hand, are expected to reduce over time based on the trend that has been taking place during the last fifteen to twenty years, and hence this variable will work in favor of V2G solutions as time goes by. Nevertheless, the cost of acquiring a vehicle usually becomes less important than the operational costs that have to be faced under a prolonged period of time, so the initial disadvantage of V2G-based solutions becomes far less significant in the long term. Furthermore, the system could be extended to public buildings and facilities, like parking lots, as long as the costs associated to that infrastructure made it worthy. Finally, according to the parameters used, the more a vehicle must be used, the more economically efficient V2G technology is. If a consumer is considering turning into a prosumer by means of a V2G solution, the mathematical model presented here is holistic enough to be applied to any other numeric values that the consumer may want to choose. If this latter idea is fully taken into account, the transition from a model made up by privately owned vehicles that are purchased and used during a very limited amount of time every day, to one based on vehicle sharing where many different users that do not own the vehicle use it almost continuously, becomes an alternative to consider.

Finally, should new challenges come up for sharing small spaces, such as the one represented by the COVID–19, sanitary protocols can be used to minimize the risk of infection to the greatest possible extent. While significant drops in car selling have taken place during the COVID–19 crisis [77], they are still usable, valid tools for transportation. Procedures used in public transport could be extended to car sharing, due to the resembling nature of all these use cases, such as (a) periodic vehicle sanitation, (b) periods of time for ventilation every timeslot a car is used, or (c) regular precaution measures that have become widespread during the pandemic (usage of disposable or washable masks and gloves) are but a few of the actions that could take place. There are several research lines that can be suggested as future works in safe usage of shared private and public means of transport, like applying COVID–19 AI-based prediction models [78] for EVV2G sharing.

Author Contributions: J.R.-M. provided the introduction, a significant part of the related works review, the foundations of the mathematical model and its numerical assessment, and the impact that the implementation of V2G would have in the power grids. P.C. made contributions to the equations of the mathematical model and the calculations derived from it, along with a review of the state of the art and the addition of some references to it. V.B. reviewed the whole manuscript and added notes to the parts that had to be improved in terms of quantity and quality. M.M.-N. participated in the overall readability of the manuscript, reviewed the mathematical model and cooperated in the creation of the graphs of the manuscript and the data tables. All authors have read and agreed to the published version of the manuscript.

Funding: This research received no external funding.

Conflicts of Interest: The authors declare no conflict of interest.

Nomenclature

Term	Meaning
Average distance rate	Used for maintenance and externalities measurement
Battery degradation	Used to calculate when the battery will have to be replaced
Battery leasing	Cost of leasing the battery
Battery replacement	Needed for continued energy storage
Capex	Cost of acquiring the vehicle
Carbon emissions	Quantity of carbon released by the vehicle
Cost of the Fuel	Evaluates difference between Internal Combustion Engine gas and electricity
Distance	Number of kilometres run by the vehicle
Energy consumption	Evaluates the consumed resources by the vehicle
Energy loses	Evaluates the loss of energy in the vehicle operation
Externalities	Impact in other areas related to the vehicle environment
Fuel consumption	Required to start and run the vehicle
Labour cost of battery change	Cost of a battery change
Labour cost of gas refilling	Cost of refilling gas
Maintenance	Costs used to keep the vehicle usable
Maximum real capacity	Actual capacity of the battery rather than the nominal one
Opex	Operational costs to keep the vehicle functional
Passive energy losses	Resulting from leaving the battery idle
Purchased energy	Electricity bought for the vehicle
Purchased power	Power bought during a certain amount of time
Revenues	Benefits from trading operations
Round-trip efficiency	Efficiency of energy usage in a full charge cycle
Social cost of carbon	Used to assess the impact on the environment by the vehicle
Sold energy	Electricity sold through the vehicle
Time	Amount of time for energy purchases
V2G conversion	Cost of turning an EV into a V2G

Appendix A

Tables and graphs representing calculations have been included in this manuscript in order to offer a clearer view of the change in costs and expenses over the amount of time used. To begin with, Table A1 shows how the calculations are done regarding the first scenario included in this manuscript, and how they increase each year. It can be seen that almost from the very beginning using a V2G vehicle is advantageous compared to an ICE one. Likewise, Table A2 shows how the costs evolve for frequent drivers during the same timespan that has been established for the other two scenarios. In a similar manner, Table A3 shows how calculations result for the use case of occasional drivers. All these figures have been calculated with and without adjusting them to inflation, so the impact of the expenditures under constant 2019 USD value and in a way that could evolve in the future can be shown. It has to be taken into account that although inflation overall has been regarded to have a similar impact on the ICE vehicle and the V2G solution, it is higher for fuel than for electricity, according to the historical data that has been retrieved. For the total costs combined, cumulative figures have been included in the two rightmost columns in each of the tables. Table A4 shows the increase in the expenditures for the V2G where the battery is either purchased or rented during the same period of time, whereas Figure 3 shows the graphical representation. It has been depicted how from the sixth year and on costs favor purchasing the battery of the vehicle instead of renting it, with the exception of the period of time taking place immediately after renewing the battery of the V2G vehicle. Finally, Tables A4–A7 show yearly difference in trading profitability depending on whether battery degradation is present or not. An example of this use case would be the yearly replacement of the battery in a V2G solution that is rented rather than purchased as part of the car.

Table A1. Cumulative costs for ICE and V2G solutions, professional drivers (figures are in USD).

CAPEX ICE	Year	Externalities ICE	Ext. ICE w. Inflation	Maintenance	Maint. w. Inflation	Fuel	Fuel w. Inflation	OPEX ICE	OPEX ICE w. Inflation	CAPEX + OPEX ICE (Cumulative)	CAPEX + OPEX ICE w. Inflation (Cumulative)
35,285											
	1	5202.59	5202.59	30,546.19	30,546.19	6121.09	6121.09	41,869.86	41,869.86	77,154.86	77,154.86
	2	5202.59	5294.16	30,546.19	31,083.80	6121.09	6353.69	41,869.86	42,731.64	119,024.73	119,886.51
	3	5202.59	5387.33	30,546.19	31,630.87	6121.09	6595.13	41,869.86	43,613.34	160,894.59	163,499.85
	4	5202.59	5482.15	30,546.19	32,187.58	6121.09	6845.74	41,869.86	44,515.47	202,764.46	208,015.32
	5	5202.59	5578.64	30,546.19	32,754.08	6121.09	7105.88	41,869.86	45,438.60	244,634.32	253,453.92
	6	5202.59	5676.82	30,546.19	33,330.55	6121.09	7375.91	41,869.86	46,383.28	286,504.19	299,837.19
	7	5202.59	5776.73	30,546.19	33,917.17	6121.09	7656.19	41,869.86	47,350.09	328,374.05	347,187.28
	8	5202.59	5878.40	30,546.19	34,514.11	6121.09	7947.13	41,869.86	48,339.64	370,243.92	395,526.92
	9	5202.59	5981.86	30,546.19	35,121.56	6121.09	8249.12	41,869.86	49,352.54	412,113.78	444,879.46
	10	5202.59	6087.14	30,546.19	35,739.70	6121.09	8562.58	41,869.86	50,389.42	453,983.65	495,268.88
	11	5202.59	6194.28	30,546.19	36,368.72	6121.09	8887.96	41,869.86	51,450.96	495,853.51	546,719.84
	12	5202.59	6303.30	30,546.19	37,008.81	6121.09	9225.70	41,869.86	52,537.81	537,723.38	599,257.65
	Total	62,431.08	68,843.39	366,554.24	404,203.13	73.453,06	90,926.12	502,438.38	563,972.65	537,723.38	599,257.65

CAPEX V2G	Year	Externalities V2G	Ext. V2G w. Inflation	Maintenance	Maint. w. Inflation	Electr.	Elec. w. Inflation	OPEX V2G	OPEX V2G w. Inflation	CAPEX + OPEX V2G + Battery (Cumulative)	CAPEX + OPEX V2G w. Inflation + Battery (Cumulative)
44,721											
	1	927.88	927.88	6868.48	6868.48	1387.16	1387,16	9183.52	9183.52	53,904.52	53,904.52
	2	927.88	944.21	6868.48	6989.36	1403.26	1429,92	9199.62	9363.49	63,104.13	63,268.01
	3	927.88	960.83	6868.48	7112.38	1419.36	1473,29	9215.72	9546,50	72,319.85	72,814.51
	4	927.88	977.74	6868.48	7237.55	1435.46	1517,28	9231.81	9732.57	81,551.66	82,547.08
	5	927.88	994.95	6868.48	7364.93	1451.55	1561,87	9247.91	9921.75	90,799.58	92,468.83
	6	927.88	1012.46	6868.48	7494.56	1467.65	1607,08	9264.01	10,114.10	100,063.59	102,582.93
	7	927.88	1030.28	6868.48	7626.46	1483.75	1652,90	9280.11	10,309.64	109,343.70	112,892.57
	8	927.88	1048.41	6868.48	7760.69	1499.85	1699,33	9296.21	10,508.43	118,639.90	123,401.00
	9	927.88	1066.86	6868.48	7897.27	1515.95	1746,37	9312.31	10,710.51	134,957.81	141,117.11
	10	927.88	1085.64	6868.48	8036.27	1532.05	1794,03	9328.41	10,915.93	144,286.22	152,033.05
	11	927.88	1104.75	6868.48	8177.71	1548.15	1842,29	9344.50	11,124.75	153,630.72	163,157.79
	12	927.88	1124.19	6868.48	8321.63	1564.25	1891,17	9360.60	11,337.00	162,991.32	174,494.79
	Total	11,134.56	12,278.20	82,421.73	90,887.29	17,708.43	19,602.71	111,264.72	122,768.19	162,991.32	174,494.79

Table A2. Cumulative costs for ICE and V2G solutions, frequent drivers (figures are in USD).

CAPEX ICE	Year	Externalities ICE	Ext. ICE w. Inflation	Maintenance	Maint. w. Inflation	Fuel	Fuel w. Inflation	OPEX ICE	OPEX ICE w. Inflation	CAPEX + OPEX ICE	CAPEX + OPEX ICE w. Inflation
35,285	1	969.19	969.19	6708.47	6708.47	1140.22	1140.22	8817.89	8817.89	44,102.89	44,102.89
	2	969.19	986.25	6708.47	6826.54	1140.22	1183.55	8817.89	8996.34	52,920.78	53,099.23
	3	969.19	1003.61	6708.47	6946.69	1140.22	1228.53	8817.89	9178.82	61,738.66	62,278.05
	4	969.19	1021.27	6708.47	7068.95	1140.22	1275.21	8817.89	9365.43	70,556.55	71,643.49
	5	969.19	1039.24	6708.47	7193.37	1140.22	1323.67	8817.89	9556.28	79,374.44	81,199.77
	6	969.19	1057.53	6708.47	7319.97	1140.22	1373.97	8817.89	9751.47	88,192.33	90,951.24
	7	969.19	1076.15	6708.47	7448.80	1140.22	1426.18	8817.89	9951.13	97,010.22	100,902.37
	8	969.19	1095.09	6708.47	7579.90	1140.22	1480.37	8817.89	10,155.36	105,828.10	111,057.73
	9	969.19	1114.36	6708.47	7713.31	1140.22	1536.63	8817.89	10,364.29	114,645.99	121,422.02
	10	969.19	1133.97	6708.47	7849.06	1140.22	1595.02	8817.89	10,578.05	123,463.88	132,000.07
	11	969.19	1153.93	6708.47	7987.20	1140.22	1655.63	8817.89	10,796.77	132,281.77	142,796,84
	12	969.19	1174.24	6708.47	8127.78	1140.22	1718.55	8817.89	11,020.56	141,099.66	153,817.40
	Total	11,630.28	12,824.83	80,501.69	88,770.04	13,682.69	16,937.54	105,814.66	118,532.40	141,099.66	153,817.40

CAPEX V2G	Year	Externalities V2G	Ext. V2G w. Inflation	Maintenance	Maint. w. Inflation	Electr.	Elec. w. Inflation	OPEX V2G	OPEX V2G w. Inflation	CAPEX + OPEX V2G + Battery	CAPEX + OPEX V2G w. Inflation + Battery
44,721	1	244.70	244.70	1559.96	1559.96	−788.42	−788.42	1016.24	1016.24	45,737.24	45,737.24
	2	244.70	249.40	1559.96	1589.91	−772.33	−787.01	1032.33	1052.30	46,769.57	46,789.55
	3	244.70	254.19	1559.96	1620.44	−756.25	−784.99	1048.41	1089.64	47,817.98	47,879.18
	4	244.70	259.07	1559.96	1651.55	−740.17	−782.36	1064.49	1128.26	55,888.07	56,013.04
	5	244.70	264.04	1559.96	1683.26	−724.08	−779.11	1080.58	1168.19	56,968.65	57,181.23
	6	244.70	269.11	1559.96	1715.58	−708.00	−775.26	1096.66	1209.43	58,065.31	58,390.66
	7	244.70	274.28	1559.96	1748.52	−691.92	−770.80	1112.74	1252.00	66,183.65	66,648.25
	8	244.70	279.54	1559.96	1782.09	−675.84	−765.72	1128.82	1295.91	67,312.47	67,944.17
	9	244.70	284.91	1559.96	1816.31	−659.75	−760.04	1144.91	1341.18	68,457.38	69,285.35
	10	244.70	290.38	1559.96	1851.18	−643.67	−753.74	1160.99	1387.82	76,623.97	77,678.77
	11	244.70	295.96	1559.96	1886.72	−627.59	−746.83	1177.07	1435.85	77,801.04	79,114.61
	12	244.70	301.64	1559.96	1922.95	−611.51	−739.31	1193.15	1485.27	78,994.20	80,599.89
	Total	2936.40	3267.21	18,719.52	20,828.45	−8399.52	−9233.58	13,256.40	14,862.09	78,994.20	80,599.89

Table A3. Cumulative costs for ICE and V2G solutions, occasional drivers (figures are in USD).

CAPEX ICE	Year	Externalities ICE	Ext. ICE w. Inflation	Maintenance	Maint. w. Inflation	Fuel	Fuel w. Inflation	OPEX ICE	OPEX ICE w. Inflation	CAPEX + OPEX ICE	CAPEX + OPEX ICE w. Inflation
35,285	1	242,30	242,30	2615.37	2615.37	285.06	285.06	3142.72	3142.72	38,427.72	38,427.72
	2	242,30	246,56	2615.37	2661.40	285.06	295.89	3142.72	3203.85	41,570.44	41,631.57
	3	242,30	250,90	2615.37	2708.24	285.06	307.13	3142.72	3266.27	44,713.17	44,897.84
	4	242,30	255,32	2615.37	2755.90	285.06	318.80	3142.72	3330.03	47,855.89	48,227.87
	5	242,30	259,81	2615.37	2804.41	285.06	330.92	3142.72	3395.14	50,998.61	51,623.01
	6	242,30	264,38	2615.37	2853.77	285.06	343.49	3142.72	3461.64	54,141.33	55,084.65
	7	242,30	269,04	2615.37	2903.99	285.06	356.54	3142.72	3529.57	57,284.06	58,614.22
	8	242,30	273,77	2615.37	2955.10	285.06	370.09	3142.72	3598.97	60,426.78	62,213.19
	9	242,30	278,59	2615.37	3007.11	285.06	384.16	3142.72	3669.86	63,569.50	65,883.05
	10	242,30	283,49	2615.37	3060.04	285.06	398.76	3142.72	3742.29	66,712.22	69,625.34
	11	242,30	288,48	2615.37	3113.89	285.06	413.91	3142.72	3816.29	69,854.94	73,441.62
	12	242,30	293,56	2615.37	3168.70	285.06	429.64	3142.72	3891.90	72,997.67	77,333.52
	Total	**2907.57**	**3206.21**	**31,384.42**	**34,607.92**	**3420.67**	**4234.38**	**37,712.67**	**42,048.52**	**72,997.67**	**77,333.52**

CAPEX V2G	Year	Externalities V2G	Ext. V2G w. Inflation	Maintenance	Maint. w. Inflation	Electr.	Elec. w. Inflation	OPEX V2G	OPEX V2G w. Inflation	CAPEX + OPEX V2G + Battery	CAPEX + OPEX V2G w. Inflation + Battery
44,721	1	60.84	60.84	1512.62	1512.62	−1157.36	−1157.36	416.11	416.11	45,137.11	45,137.11
	2	60.84	61.91	1512.62	1539.25	−1141.33	−1163.02	432.13	438.14	45,569.24	45,575.24
	3	60.84	63.00	1512.62	1566.34	−1125.31	−1168.07	448.15	461.26	46,017.39	46,036.51
	4	60.84	64.11	1512.62	1593.90	−1109.29	−1172.52	464.18	485.50	46,481.57	46,522.01
	5	60.84	65.24	1512.62	1621.96	−1093.26	−1176.35	480.20	510.84	46,961.77	47,032.85
	6	60.84	66.39	1512.62	1650.50	−1077.24	−1179.58	496.22	537.31	47,457.99	47,570.16
	7	60.84	67.55	1512.62	1679.55	−1061.22	−1182.20	512.25	564.91	47,970.24	48,135.07
	8	60.84	68.74	1512.62	1709.11	−1045.19	−1184.20	528.27	593.65	48,498.51	48,728.72
	9	60.84	69.95	1512.62	1739.19	−1029.17	−1185.60	544.29	623.54	56,048.40	56,357.86
	10	60.84	71.18	1512.62	1769.80	−1013.15	−1186.39	560.32	654.59	56,608.72	57,012.46
	11	60.84	72.44	1512.62	1800.95	−997.12	−1186.58	576.34	686.81	57,185.06	57,699.27
	12	60.84	73.71	1512.62	1832.65	−981.10	−1186.15	592.36	720.21	57,777.42	58,419.48
	Total	**730.08**	**805.07**	**18,151.48**	**20,015.83**	**−12,830.74**	**−14,128.02**	**6050.82**	**6692.88**	**57,777.42**	**58,419.48**

Table A4. Cumulative costs for battery rental and battery purchase for the V2G solution (figures are in USD).

Year	CAPEX + Battery Rental	CAPEX + Battery Rental+ Inflation	V2G Battery Purchase	Cost Difference Rental vs. Purchase	Cost Difference Rental vs. Purchase + Inflation
1	35,285	35,285	44,721	−9436.00	−9436.00
2	37,715.40	36,994.57	44,721	−7005.60	−7726.43
3	39,395.40	39,455.06	44,721	−5325.60	−5265.94
4	41,075.40	41,165.67	44,721	−3645.60	−3555.33
5	42,755.40	42,876.83	44,721	−1965.60	−1844.17
6	44,435.40	44,588.54	44,721	−285.60	−132.46
7	46,115.40	46,300.80	44,721	1394.40	1579.80
8	47,795.40	48,013.63	44,721	3074.40	3292.63
9	49,475.40	49,727.04	51,726.60	−2251.20	−1999.56
10	51,155.40	51,441.04	51,726.60	−571.20	−285.56
11	52,835.40	53,155.63	51,726.60	1108.80	1429.03
12	54,515.40	54,870.84	51,726.60	2788.80	3144.24

Table A5. Costs/profits with and without yearly degradation (professional drivers).

Year	Bought Energy	Cost per KWh Bought	Cost per KWh + Inflation	Sold Energy	Cost per KWh Sold	Cost per KWh Sold + Inflation	Energy V2G	Energy V2G + Inflation
1	40,000	0.09956	0.09956	19,695.970	0.131765	0.131765	1387.16	1387,16
2	39,500	0.09956	0.10145164	19,196.000	0.131765	0.134268535	1403.26	1429,92
3	39,000	0.09956	0.10334328	18,696.029	0.131765	0.13677207	1419.36	1473,29
4	38,500	0.09956	0.10523492	18,196.059	0.131765	0139275605	1435.46	1517,28
5	38,000	0.09956	0.10712656	17,696.088	0.131765	0.14177914	1451.55	1561,87
6	37,500	0.09956	0.1090182	17,196.118	0.131765	0.144282675	1467.65	1607,08
7	37,000	0.09956	0.11090984	16,696.147	0.131765	0.14678621	1483.75	1652,90
8	36,500	0.09956	0.11280148	16,196.177	0.131765	0.149289745	1499.85	1699,33
9	36,000	0.09956	0.11469312	15,696.206	0.131765	0.15179328	1515.95	1746,37
10	35,500	0.09956	0.11658476	15,196.236	0.131765	0.154296815	1532.05	1794,03
11	35,000	0.09956	0.1184764	14,696.265	0.131765	0.15680035	1548.15	1842,29
12	34,500	0.09956	0.12036804	14,196.295	0.131765	0.159303885	1564.25	1891,17
1	40,000	0.09956	0.09956	19,695.97	0.131765	0.131765	1387.16	1387.16
2	40,000	0.09956	0.10145164	19,695.97	0.131765	0.134268535	1387.16	1413.52
3	40,000	0.09956	0.10334328	19,695.97	0.131765	0.13677207	1387.16	1439.87
4	40,000	0.09956	0.10523492	19,695.97	0.131765	0.139275605	1387.16	1466.23
5	40,000	0.09956	0.10712656	19,695.97	0.131765	0.14177914	1387.16	1492.58
6	40,000	0.09956	0.1090182	19,695.97	0.131765	0.144282675	1387.16	1518.94
7	40,000	0.09956	0.11090984	19,695.97	0.131765	0.14678621	1387.16	1545.30
8	40,000	0.09956	0.11280148	19,695.97	0.131765	0.149289745	1387.16	1571.65
9	40,000	0.09956	0.11469312	19,695.97	0.131765	0.15179328	1387.16	1598.01
10	40,000	0.09956	0.11658476	19,695.97	0.131765	0.154296815	1387.16	1624.36
11	40,000	0.09956	0.1184764	19,695.97	0.131765	0.15680035	1387.16	1650.72
12	40,000	0.09956	0.12036804	19,695.97	0.131765	0.159303885	1387.16	1677.08

Table **A6.** Energy costs/profits with and without yearly degradation (frequent drivers).

Year	Bought Energy	Cost per KWh Bought	Cost per KWh + Inflation	Sold Energy	Cost per KWh Sold	Cost per KWh Sold + Inflation	Energy V2G	Energy V2G + Inflation
1	40,000	0.09956	0.09956	36,207.000	0.131765	0.131765	−788.42	−788.42
2	39,500	0.09956	0.10145164	35,707.150	0.131765	0.134268535	−772.33	−787.01
3	39,000	0.09956	0.10334328	35,207.300	0.131765	0.13677207	−756.25	−784.99
4	38,500	0.09956	0.10523492	34,707.450	0.131765	0139275605	−740.17	−782.36
5	38,000	0.09956	0.10712656	34,207.600	0.131765	0.14177914	−724.08	−779.11
6	37,500	0.09956	0.1090182	33,707.750	0.131765	0.144282675	−708.00	−775.26
7	37,000	0.09956	0.11090984	33,207.900	0.131765	0.14678621	−691.92	−770.80
8	36,500	0.09956	0.11280148	32,708.050	0.131765	0.149289745	−675.84	−765.72
9	36,000	0.09956	0.11469312	32,208.200	0.131765	0.15179328	−659.75	−760.04
10	35,500	0.09956	0.11658476	31,708.350	0.131765	0.154296815	−643.67	−753.74
11	35,000	0.09956	0.1184764	31,208.500	0.131765	0.15680035	−627.59	−746.83
12	34,500	0.09956	0.12036804	30,708.650	0.131765	0.159303885	−611.51	−739.31
1	40,000	0.09956	0.09956	36,207.00	0.131765	0.131765	−788.42	−788.42
2	40,000	0.09956	0.10145164	36,207.00	0.131765	0.134268535	−788.42	−803.40
3	40,000	0.09956	0.10334328	36,207.00	0.131765	0.13677207	−788.42	−818.38
4	40,000	0.09956	0.10523492	36,207.00	0.131765	0.139275605	−788.42	−833.36
5	40,000	0.09956	0.10712656	36,207.00	0.131765	0.14177914	−788.42	−848.33
6	40,000	0.09956	0.1090182	36,207.00	0.131765	0.144282675	−788.42	−863.31
7	40,000	0.09956	0.11090984	36,207.00	0.131765	0.14678621	−788.42	−878.29
8	40,000	0.09956	0.11280148	36,207.00	0.131765	0.149289745	−788.42	−893.27
9	40,000	0.09956	0.11469312	36,207.00	0.131765	0.15179328	−788.42	−908.25
10	40,000	0.09956	0.11658476	36,207.00	0.131765	0.154296815	−788.42	−923.23
11	40,000	0.09956	0.1184764	36,207.00	0.131765	0.15680035	−788.42	−938.21
12	40,000	0.09956	0.12036804	36,207.00	0.131765	0.159303885	−788.42	−953.19

Table A7. Energy costs/profits with and without yearly degradation (occasional drivers).

Year	Bought Energy	Cost per KWh Bought	Cost per KWh + Inflation	Sold Energy	Cost per KWh Sold	Cost per KWh Sold + Inflation	Energy V2G	Energy V2G + Inflation
1	40,000	0.09956	0.09956	39,007.000	0.131765	0.131765	−1157.36	−1157.36
2	39,500	0.09956	0.10145164	38,507.600	0.131765	0.134268535	−1141.33	−1163.02
3	39,000	0.09956	0.10334328	38,008.200	0.131765	0.13677207	−1125.31	−1168.07
4	38,500	0.09956	0.10523492	37,508.800	0.131765	0139275605	−1109.29	−1172.52
5	38,000	0.09956	0.10712656	37,009.400	0.131765	0.14177914	−1093.26	−1176.35
6	37,500	0.09956	0.1090182	36,510.000	0.131765	0.144282675	−1077.24	−1179.58
7	37,000	0.09956	0.11090984	36,010.600	0.131765	0.14678621	−1061.22	−1182.20
8	36,500	0.09956	0.11280148	35,511.200	0.131765	0.149289745	−1045.19	−1184.20
9	36,000	0.09956	0.11469312	35,011.800	0.131765	0.15179328	−1029.17	−1185.60
10	35,500	0.09956	0.11658476	34,512.400	0.131765	0.154296815	−1013.15	−1186.39
11	35,000	0.09956	0.1184764	34,013.000	0.131765	0.15680035	−997.12	−1186.58
12	34,500	0.09956	0.12036804	33,513.600	0.131765	0.159303885	−981.10	−1186.15
1	40,000	0.09956	0.09956	39,007.00	0.131765	0.131765	−1157.36	−1157.36
2	40,000	0.09956	0.10145164	39,007.00	0.131765	0.134268535	−1157.36	−1179.35
3	40,000	0.09956	0.10334328	39,007.00	0.131765	0.13677207	−1157.36	−1201.34
4	40,000	0.09956	0.10523492	39,007.00	0.131765	0.139275605	−1157.36	−1223.33
5	40,000	0.09956	0.10712656	39,007.00	0.131765	0.14177914	−1157.36	−1245.32
6	40,000	0.09956	0.1090182	39,007.00	0.131765	0.144282675	−1157.36	−1267.31
7	40,000	0.09956	0.11090984	39,007.00	0.131765	0.14678621	−1157.36	−1289.30
8	40,000	0.09956	0.11280148	39,007.00	0.131765	0.149289745	−1157.36	−1311.29
9	40,000	0.09956	0.11469312	39,007.00	0.131765	0.15179328	−1157.36	−1333.28
10	40,000	0.09956	0.11658476	39,007.00	0.131765	0.154296815	−1157.36	−1355.27
11	40,000	0.09956	0.1184764	39,007.00	0.131765	0.15680035	−1157.36	−1377.26
12	40,000	0.09956	0.12036804	39,007.00	0.131765	0.159303885	−1157.36	−1399.25

References

1. Gungor, V.C.; Sahin, D.; Kocak, T.; Ergut, S.; Buccella, C.; Cecati, C.; Hancke, G.P. Smart Grid Technologies: Communication Technologies and Standards. *IEEE Trans. Ind. Inform.* **2011**, *7*, 529–539. [CrossRef]
2. Chen, X.; Dinh, H.; Wang, B. Cascading Failures in Smart Grid—Benefits of Distributed Generation. In Proceedings of the First IEEE International Conference on Smart Grid Communications, Gaithersburg, MD, USA, 4–6 October 2010.
3. Internal Revenue Service. Plug-In Electric Drive Vehicle Credit (IRC 30D). Available online: https://www.irs.gov/businesses/plug-in-electric-vehicle-credit-irc\$-\$30-and-irc\$-\$30d (accessed on 15 September 2020).
4. Sakurama, K.; Miura, M. Real-time pricing via distributed negotiations between prosumers in smart grids. In Proceedings of the IEEE Innovative Smart Grid Technologies—Asia (ISGT ASIA), Bangkok, Thailand, 3–6 November 2015.
5. Palensky, P.; Dietrich, D. Demand Side Management: Demand Response, Intelligent Energy Systems, and Smart Loads. *IEEE Trans. Ind. Inform.* **2011**, *7*, 381–388. [CrossRef]
6. United States Department of Energy. Emissions from Hybrid and Plug-In Electric Vehicles. Available online: http://www.afdc.energy.gov/vehicles/electric_emissions.php (accessed on 15 September 2020).
7. Gungor, V.C.; Sahin, D.; Kocak, T.; Ergut, S.; Buccella, C.; Cecati, C.; Hancke, G.P. A Survey on Smart Grid Potential Applications and Communication Requirements. *IEEE Trans. Ind. Inform.* **2013**, *9*, 28–42. [CrossRef]
8. Cai, H.; Hu, G. Distributed Control Scheme for Package-Level State-of-Charge Balancing of Grid-Connected Battery Energy Storage System. *IEEE Trans. Ind. Inform.* **2016**, *12*, 1919–1929. [CrossRef]
9. Noel, L.; McCormack, R. A cost benefit analysis of a V2G-capable electric school bus compared to a traditional diesel school bus. *Appl. Energy* **2014**, *126*, 246–255. [CrossRef]
10. Shirazi, Y.; Carr, E.D.; Knapp, L. A cost–benefit analysis of alternatively fueled buses with special considerations for V2G technology. *Energy Policy* **2015**, *87*, 591–603. [CrossRef]
11. Park, D.; Seungwook, Y.; Euiseok, H. Cost benefit analysis of public service electric vehicles with vehicle-to-grid (V2G) capability. In Proceedings of the IEEE Transportation Electrification Conference and Expo, Asia-Pacific (ITEC Asia-Pacific), Busan, Korea, 1–4 June 2016.
12. Nworgu, O.A.; Chukwu, U.C.; Okezie, C.G.; Chukwu, N.B. Economic prospects and market operations of V2G in electric distribution network. In Proceedings of the IEEE/PES Transmission and Distribution Conference and Exposition (T&D), Dallas, TX, USA, 3–5 May 2016.
13. Hill, D.M.; Agarwal, A.S.; Ayello, F. Fleet operator risks for using fleets for V2G regulation. *Energy Policy* **2012**, *41*, 221–231. [CrossRef]
14. Musio, M.; Lombardi, P.; Damiano, A. Vehicles to grid (V2G) concept applied to a Virtual Power Plant structure. In Proceedings of the XIX International Conference on Electrical Machines—ICEM, Rome, Italy, 6–8 September 2010.
15. Jain, P.; Meena, D.; Jain, T. Revenue valuation of aggregated electric vehicles participating in V2G power service. In Proceedings of the IEEE Innovative Smart Grid Technologies—Asia (ISGT ASIA), Bangkok, Thailand, 3–6 November 2015.
16. Lund, H.; Kempton, W. Integration of renewable energy into the transport and electricity sectors through V2G. *Energy Policy* **2008**, *36*, 3578–3587. [CrossRef]
17. Qiang, H.; Gu, Y.; Zheng, J.; Zhou, X. Modeling and Simulating of Private EVs Charging Load. *Open Electr. Electron. Eng. J.* **2015**, *9*, 231–237. [CrossRef]
18. Kumar, S.; Yaragatti, U.R.; Manasani, S. Modeling and Architectural Frame Work of Off-Board V2G Integrator for Smart Grid. *Int. J. Renew. Energy Res.* **2014**, *4*, 826–831.
19. Du, C.; He, J. V2G charge-discharge strategy with EV mass application. In Proceedings of the 21st International Conference on Electricity Distribution, Franfurt, Germany, 6–9 June 2011.
20. Wang, Z.; Tang, Y.; Chen, X.; Men, X.; Cao, J.; Wang, H. Optimized Daily Dispatching Strategy of Building—Integrated Energy Systems Considering Vehicle to Grid Technology and Room Temperature Control. *Energies* **2018**, *11*, 1287. [CrossRef]
21. Li, Y.; Zhang, P.; Wang, Y. The Location Privacy Protection of Electric Vehicles with Differential Privacy in V2G Networks. *Energies* **2018**, *11*, 2625. [CrossRef]
22. Harighi, T.; Bayindir, R.; Padmanaban, S.; Mihet-Popa, L.; Hossain, E. An Overview of Energy Scenarios, Storage Systems and the Infrastructure for Vehicle-to-Grid Technology. *Energies* **2018**, *11*, 2174. [CrossRef]

23. Child, M.; Nordling, A.; Breyer, C. The Impacts of High V2G Participation in a 100% Renewable Åland Energy System. *Energies* **2018**, *11*, 2206. [CrossRef]

24. Weldon, P.; Morrissey, P.; O'Mahony, M. Long-term cost of ownership comparative analysis between electric vehicles and internal combustion engine vehicles. *Sustain. Cities Soc.* **2018**, *39*, 578–591. [CrossRef]

25. Zhang, Y.; Lu, M.; Shen, S. On the Values of Vehicle-to-Grid Electricity Selling in Electric Vehicle Sharing. 2018. Available online: https://ssrn.com/abstract=3172116 (accessed on 18 September 2020).

26. Maeng, K.; Ko, S.; Shin, J.; Cho, Y. How Much Electricity Sharing Will Electric Vehicle Owners Allow from Their Battery? Incorporating Vehicle-to-Grid Technology and Electricity Generation Mix. *Energies* **2020**, *13*, 4248. [CrossRef]

27. Ercan, T.; Noori, M.; Zhao, Y.; Tatari, O. On the Front Lines of a Sustainable Transportation Fleet: Applications of Vehicle-to-Grid Technology for Transit and School Buses. *Energies* **2016**, *9*, 230. [CrossRef]

28. Noel, L.; Brodie, J.F.; Kempton, W.; Archer, C.L.; Budischak, C. Cost minimization of generation, storage, and new loads, comparing costs with and without externalities. *Appl. Energy* **2017**, *189*, 110–121. [CrossRef]

29. Erdinc, O.; Mendes, T.D.P.; Catalão, J.P.S. Impact of electric vehicle V2G operation and demand response strategies for smart households. In Proceedings of the IEEE PES T&D Conference and Exposition, Chicago, IL, USA, 14–17 April 2014; pp. 1–5. [CrossRef]

30. Zeng, M.; Leng, S.; Maharjan, S.; Gjessing, S.; He, J. An Incentivized Auction-Based Group-Selling Approach for Demand Response Management in V2G Systems. *IEEE Trans. Ind. Inform.* **2015**, *11*, 1554–1563. [CrossRef]

31. Kumar, K.N.; Sivaneasan, B.; Cheah, P.H.; So, P.L.; Wang, D.Z.W. V2G Capacity Estimation Using Dynamic EV Scheduling. *IEEE Trans. Smart Grid* **2014**, *5*, 1051–1060. [CrossRef]

32. You, S.; Hu, J.; Pedersen, A.B.; Andersen, P.B.; Rasmussen, C.N.; Cha, S. Numerical comparison of optimal charging schemes for Electric Vehicles. In Proceedings of the IEEE Power and Energy Society General Meeting, San Diego, CA, USA, 22–26 July 2012; pp. 1–6. [CrossRef]

33. Kumar, S.; Kumar, R.Y.U. Performance analysis of LTE protocol for EV to EV communication in vehicle-to-grid (V2G). In Proceedings of the IEEE 28th Canadian Conference on Electrical and Computer Engineering (CCECE), Halifax, NS, Canada, 3–6 May 2015; pp. 1567–1571. [CrossRef]

34. Parsons, G.R.; Hidrue, M.K.; Kempton, W.; Gardner, M.P. Willingness to Pay for Vehicle-to-Grid (V2G) Electric Vehicles and Their Contract Terms. *Energy Econ.* **2014**, *42*. [CrossRef]

35. Red Eléctrica de España Web Site. Active Energy Invoicing Price. Available online: https://www.esios.ree.es/en/pvpc (accessed on 15 September 2020).

36. Paevere, P.; Higgins, A.; Ren, Z.; Horn, M.; Grozev, G.; McNamara, C. Spatio-temporal modelling of electric vehicle charging demand and impacts on peak household electrical load. *Sustain. Sci.* **2014**, *9*, 10–1007. [CrossRef]

37. Ashtari, A.; Bibeau, E.; Shahidinejad, S.; Molinski, T. PEV Charging Profile Prediction and Analysis Based on Vehicle Usage Data. *IEEE Trans. Smart Grid* **2012**, *3*, 341–350. [CrossRef]

38. Red Eléctrica de España Web Site. Default Tariff of Active Energy Invoicing Price. Available online: https://www.esios.ree.es/en/analysis/1013 (accessed on 15 September 2020).

39. Wang, D.; Saxena, S.; Coignard, J.; Iosifidou, E.A.; Guan, X. Quantifying electric vehicle battery degradation from driving vs. V2G services. In Proceedings of the IEEE Power and Energy Society General Meeting (PESGM), Boston, MA, USA, 17–21 July 2016.

40. Ribberink, H.; Darcovich, K.; Pincet, F. Battery Life Impact of Vehicle-to-Grid Application of Electric Vehicles. In Proceedings of the 28th International Electric Vehicle Symposium and Exhibition, Goyang, Korea, 3–6 May 2015; p. 11.

41. Tesla Battery Range Data. Available online: https://docs.google.com/spreadsheets/d/t024bMoRiDPIDialGnuKPsg/edit#gid=154312675 (accessed on 15 September 2020).

42. Aditya, J.P.; Ferdowsi, M. Comparison of NiMH and Li-ion batteries in automotive applications. In Proceedings of the IEEE Vehicle Power and Propulsion Conference, Harbin, China, 3–5 September 2008; pp. 1–6. [CrossRef]

43. Tesla, Inc. Vehicle Warranty, Model 3. Tesla Official Web Site. Available online: https://www.tesla.com/support/vehicle-warranty (accessed on 15 September 2020).

44. Nissan Motor Company. Warranty Information Booklet from Nissan Leaf. 2018. Available online: https://owners.nissanusa.com/content/techpub/ManualsAndGuides/LEAF/2018/2018-LEAF-warranty-booklet.pdf (accessed on 15 September 2020).

45. Palmer, K.; Tate, J.; Wadud, Z.; Nellthorp, J. Total cost of ownership and market share for hybrid and electric vehicles in the UK, US and Japan. *Appl. Energy* **2018**, *209*, 108–119. [CrossRef]

46. eFile. Tax Credits for Buying Alternative, Electric Motor Vehicles. Available online: https://www.efile.com/tax-credit/hybrid-car-tax-credit/ (accessed on 4 November 2020).

47. Hanley, S. Nissan LEAF Replacement Battery Cost = $5499. Available online: https://cleantechnica.com/2017/10/04/nissan-leaf-replacement-battery-will-cost$-$5499/ (accessed on 15 September 2020).

48. Union of Concerned Scientists. Accelerating US Leadership in Electric Vehicles. Available online: https://www.ucsusa.org/sites/default/files/attach/2017/09/cv-factsheets-ev-incentives.pdf?_ga=2.108452610.1630188791.1517413160$-$1434713090.1436805699 (accessed on 15 September 2020).

49. Berckmans, G.; Messagie, M.; Smekens, J.; Omar, N.; Vanhaverbeke, L.; Van Mierlo, J. Cost Projection of State of the Art Lithium-Ion Batteries for Electric Vehicles up to 2030. *Energies* **2017**, *10*, 1314. [CrossRef]

50. Taxi and Limousine Commission. Taxi Cab Factbook. 2014. Available online: https://www1.nyc.gov/assets/tlc/downloads/pdf/2014_tlc_factbook.pdf (accessed on 15 September 2020).

51. Suwannee Valley Electric Cooperative, Inc. Index of Rate Schedules. 2015. Available online: https://svec-coop.com/wp-content/uploads/SVEC$-$6_9_17-RATES.pdf (accessed on 15 September 2020).

52. Wilson, J. Electric Car Battery Leasing: Should I Lease or Buy the Batteries? Available online: https://www.buyacar.co.uk/cars/1523/electric-car-battery-leasing-should-i-lease-or-buy-the-batteries (accessed on 15 September 2020).

53. Kelley Blue Book. New-Car Transaction Prices up 2 Percent in March 2016, Along with Increases in Incentive Spend. Available online: https://mediaroom.kbb.com/average-new-car-prices-jump$-$2-percent-march$-$2018-suv-sales-strength-according-to-kelley-blue-book (accessed on 15 September 2020).

54. US Energy Information Administration. Carbon Dioxide Emissions Coefficients. Available online: https://www.eia.gov/environment/emissions/co2_vol_mass.php (accessed on 15 September 2020).

55. The International Council for Clean Transportation. Effects of Battery Manufacturing on Electric Vehicle Life-Cycle Greenhouse Gas Emissions. Available online: www.theicct.org (accessed on 15 September 2020).

56. AAA Gas Prices. Available online: http://gasprices.aaa.com/news/ (accessed on 19 August 2020).

57. Energy, U.D.O. Federal Tax Credits for All-Electric and Plug-in Hybrid Vehicles. Available online: https://www.fueleconomy.gov/feg/taxevb.shtml (accessed on 15 September 2020).

58. US Department of Transportation. Annual Vehicle Distance Travelled in Miles and Related Data by Highway Category and Vehicle Type. 2016. Available online: https://www.fhwa.dot.gov/policyinformation/statistics/2016/pdf/vm1.pdf (accessed on 15 September 2020).

59. AAA News Room. Your Driving Costs. How Much Are You Really Paying to Drive? Available online: https://newsroom.aaa.com/auto/your-driving-costs/ (accessed on 15 September 2020).

60. Fixr. Home Electric Vehicle Charging Station Cost. Available online: https://www.fixr.com/costs/home-electric-vehicle-charging-station (accessed on 15 September 2020).

61. US Department of Energy. Find a Car/Compare Side-by-Side. Available online: https://www.fueleconomy.gov/feg/Find.do?action=sbs&id=37066&id=37067&id=34918&id=34699 (accessed on 15 September 2020).

62. Consumer Price Index All Urban Consumers (Current Series). 12-Month Percent Change Series. Available online: https://data.bls.gov/timeseries/CUUR0000SA0?output_view=pct_12mths (accessed on 15 September 2020).

63. Marques, L.; Vasconcelos, V.; Pereirinha, P.G.; Trovão, J.P. Lithium Modular Battery Bank for Electric Vehicles. 2011. Available online: http://www.uc.pt/en/efs/research/EESEVS/f/XIICLEEE_1846_MarquesEtAl.pdf (accessed on 15 September 2020).

64. Interagency Working Group on Social Cost of Greenhouse Gases, United States Government. Technical Support Document: Technical Update of the Social Cost of Carbon for Regulatory Impact Analysis Under Executive Order 12866. 2016. Available online: https://www.epa.gov/sites/production/files/2016$-$12/documents/sc_co2_tsd_august_2016.pdf (accessed on 15 September 2020).

65. Ramsdale, J. How Much Will V2G Cost and When, Where and Who Will Make It Happen? Available online: http://www.yougen.co.uk/blog-entry/2688/How+much+will+V2G+cost+and+when\T1\textquoteright2C+where+and+who+will+make+it+happen\T1\textquoteright3F/ (accessed on 15 September 2020).

66. Rue One. Investment Summary—American Gas & Technology. Available online: https://www.rueone.com/images/marketing/RueOne-AGT-Overview.pdf (accessed on 16 September 2020).

67. Automotive News. Average U.S. Gasoline Usage Lowest in 3 Decades, Study Says. Available online: https://www.autonews.com/article/20150325/OEM06/150329911/average-u-s-gasoline-usage-lowest-in$-$3-decades-study-says (accessed on 16 September 2020).

68. Annelena Lobb. Premium Prices for Teen-Age Drivers. 2002. Available online: http://money.cnn.com/2002/03/21/pf/insurance/q_teenagers/index.htm (accessed on 15 September 2020).
69. Wright, P. Car Insurance for Occasional Drivers. Available online: http://surebuycarinsurance.com/occasional-driver-insurance/ (accessed on 15 September 2020).
70. Chukwu, U.C.; Nworgu, O.A.; Dike, D.O. Impact of V2G penetration on distribution system components using diversity factor. In Proceedings of the IEEE Southeastcon, Lexington, KY, USA, 13–16 March 2014.
71. Chukwu, U.C.; Mahajan, S.M. Impact of V2G on Distribution Feeder: An Energy Loss Reduction Approach. *Sci. Educ.* **2014**, *2*, 19–27.
72. Mahmud, K.; Morsalin, S.; Kafle, Y.R.; Town, G.E. Improved peak shaving in grid-connected domestic power systems combining photovoltaic generation, battery storage, and V2G-capable electric vehicle. In Proceedings of the IEEE International Conference on Power System Technology (POWERCON), Wollongong, NSW, Australia, 28 September–1 October 2016.
73. Wang, Z.; Wang, S. Grid Power Peak Shaving and Valley Filling Using Vehicle-to-Grid Systems. *IEEE Trans. Power Deliv.* **2013**, *28*, 1822–1829. [CrossRef]
74. Wu, Q.; Cheng, L.; Pineau, U.; Nielsen, A.H.; Østergaard, J. Impact and cost evaluation of electric vehicle integration on medium voltage distribution networks. In Proceedings of the IEEE PES Asia-Pacific Power and Energy Engineering Conference (APPEEC), Kowloon, Hong Kong, China, 8–11 December 2013; pp. 1–5.
75. Green, R.C.; Wang, L.; Alam, M. The impact of plug-in hybrid electric vehicles on distribution networks: A review and outlook. In Proceedings of the IEEE PES General Meeting, Minneapolis, MN, USA, 25–29 July 2010; pp. 1–8.
76. Rodríguez-Molina, J.; Kammen, D. Middleware Architectures for the Smart Grid: A Survey on the State-of-the-Art, Taxonomy and Main Open Issues. *IEEE Commun. Surv. Tutor.* **2018**, *20*, 2992–3033. [CrossRef]
77. Chamola, V.; Hassija, V.; Gupta, V.; Guizani, M. A Comprehensive Review of the COVID–19 Pandemic and the Role of IoT, Drones, AI, Blockchain, and 5G in Managing its Impact. *IEEE Access* **2020**, *8*, 90225–90265. [CrossRef]
78. Zheng, N.; Du, S.; Wang, J.; Zhang, H.; Cui, W.; Kang, Z.; Yang, T.; Lou, B.; Chi, Y.C.; Long, H.; et al. Predicting COVID–19 in China Using Hybrid AI Model. *IEEE Trans. Cybern.* **2020**, *50*, 2891–2904. [CrossRef] [PubMed]

Publisher's Note: MDPI stays neutral with regard to jurisdictional claims in published maps and institutional affiliations.

Article

Environmental Impacts of Integrated Photovoltaic Modules in Light Utility Electric Vehicles

Olga Kanz [1,*], Angèle Reinders [2], Johanna May [3] and Kaining Ding [1]

1   IEK-5 Photovoltaik, Forschungszentrum Jülich GmbH, 52425 Jülich, Germany; k.ding@fz-juelich.de
2   Energy Technology Group, Eindhoven University of Technology, 5600 MB Eindhoven, The Netherlands; a.h.m.e.reinders@tue.nl
3   Institute of Electrical Power Engineering (IET), Cologne University of Applied Sciences, 50678 Cologne, Germany; johanna.may@th-koeln.de
*   Correspondence: o.kanz@fz-juelich.de; Tel.: +49-2461-61-1636

Received: 20 August 2020; Accepted: 29 September 2020; Published: 1 October 2020

Abstract: This paper presents a life cycle assessment (LCA) of photovoltaic (PV) solar modules which have been integrated into electric vehicle applications, also called vehicle integrated photovoltaics (VIPV). The LCA was executed by means of GaBi LCA software with Ecoinvent v2.2 as a background database, with a focus on the global warming potential (GWP). A light utility electric vehicle (LUV) named StreetScooter Work L, with a PV array of 930 Wp, was analyzed for the location of Cologne, Germany. An operation time of 8 years and an average shadowing factor of 30% were assumed. The functional unit of this LCA is 1 kWh of generated PV electricity on-board, for which an emission factor of 0.357 kg CO_2-eq/kWh was calculated, whereas the average grid emissions would be 0.435 kg CO_2-eq/kWh. Hence, charging by PV power hence causes lower emissions than charging an EV by the grid. The study further shows how changes in the shadowing factor, operation time, and other aspects affect vehicle's emissions. The ecological benefit of charging by PV modules as compared to grid charging is negated when the shadowing factor exceeds 40% and hence exceeds emissions of 0.435 kg CO_2-eq/kWh. However, if the operation time of a vehicle with integrated PV is prolonged to 12 years, emissions of the functional unit go down to 0.221 kg CO_2-eq/kWh. It is relevant to point out that the outcomes of the LCA study strongly depend on the location of use of the vehicle, the annual irradiation, and the carbon footprint of the grid on that location.

Keywords: life cycle assessment; CO_2 emissions; photovoltaic systems; electric vehicles; VIPV

1. Introduction

The European Union (EU) has agreed on a range of policies aiming to reduce greenhouse gas emissions in various sectors of society. Since transport largely contributes to these emissions by a share of 27% of the EU's total emissions in 2016, these emissions have to be reduced. For the year 2030, this policy implies that in the EU fleet-wide CO_2 emissions of passenger cars should be reduced by 37.5% as compared to 1990 levels. For new vans and trucks, the emissions should be reduced by 31% [1]. Therefore, new strict targets require the reduction of average CO_2 emissions of new vehicles that will enter the market. Consequently, the year 2020 is widely expected to bring dramatic changes to the automotive market. Due to the aforementioned targets, manufacturers are forced to invest intensively in innovative technologies of sustainable mobility. Therefore, many automotive players focus on battery electric vehicles (BEVs). In recent years, a large number of environmental impact studies were published, analyzing the potential environmental benefits of electric vehicles (EVs). The overall conclusion is that BEVs are preferable over petrol and diesel vehicles, however only if charged by renewable energy [2]. A possible solution is to charge these cars with low-emission renewable energy technologies such as photovoltaic systems. This could be achieved by charging stations which are powered by PV systems or by photovoltaic solar modules which are built in a car's body parts, also called vehicle integrated photovoltaics (VIPV). Some vehicle manufacturers already aim at integrating PV cells in body parts of their passenger cars. One of the most recent solar powered electric vehicles is the Lightyear One of the Dutch company Lightyear. The vehicle has an integrated silicon PV array of more than 5 m^2 with a nominal installed powered of 1250 Wp. Similarly, Munich-based producer Sono Motors is planning on launching their solar electric vehicle, named Sion. The Sion's PV array has a nominal power of 1200 Wp. Solar charging in summer can add 34 km to the drive range of 255 km. With Audi´s e-tron Quattro with a nominal PV power of 400 Wp and the Toyota Prius P with a PV array of 860 Wp, two of the car industry's major players recently entered the market as well. Especially for light utility electric vehicles (LUV), VIPV could be an attractive feature due to their predictability of utilization, in particular their moments of use and daily travel distances, and their significantly larger and flat roof surface which, if covered by solar cells, can potentially yield sufficient amounts of solar power. LUVs are usually vehicles with a gross vehicle weight of no more than 3.5 metric tonnes and are optimized to be tough-built, have low operating costs, and to be used in intra-city operations. Though prior studies have often indicated that VIPV will result in lower CO_2 emissions, actual life cycle assessments (LCAs) of VIPV are barely available, and most claims until now have not been quantified or validated for the specific situation of VIPV of LUVs [3–5]. Thus, the goal of this work is to analyze how PV-powered vehicles can contribute to sustainable mobility. Therefore, an LCA focused on determining the CO_2 emissions of a German VIPV LUV called StreetScooter will be conducted. The results of this research could be useful for car manufacturers, to calculate emissions per vehicle, for political institutions to estimate environmental impacts for the transport sector, and for business parties in the solar market to identify further application possibilities and yield useful data to identify critical areas for the improvement of VIPV for LUVs.

This LCA study was executed in the framework of a project called STREET, which was funded by the German Ministry for Economic Affairs and Energy and a German logistics company of Deutsche Post DHL Group named StreetScooter, which is currently working on the integration of PV on electric light utility vehicles (see Figure 1). Forschungszentrum Jülich as an organization for applied research supports the project by equipping the vehicle with PV modules and by analyzing the energy yield of PV modules on this vehicle by the analysis of data measured by radiation sensors on the vehicle under real shading and reflection conditions.

Figure 1. StreetScooter Work L Reprinted from: CC-BY-SA-4.0 (via Wikimedia Commons), Superbass, 2017.

This paper is structured as following: in Section 2 the LCA method will be explained and all input parameters of the LCA will be described. Major assumptions regarding the operation phase and technology choice for the on-board vehicle application are discussed. The results, sensitivity analyses, and limitations of the study are reported in Section 3. Finally, Section 4 summarizes the results, presents the conclusions, and offers recommendations for future studies.

2. Method and Data

This section presents the general methodology used to execute the LCA, defines the efficiency of the VIPV investigated, and quantifies the resulting CO_2 emissions. Additionally, key parameters that limit the environmental performance of the electricity produced by the PV system integrated into the vehicle are shown. Assumptions about these critical parameters for the reference case are clarified.

2.1. Life Cycle Assessment Method

LCA is a useful tool to quantify environmental performance, considering a holistic perspective. LCA is generally understood as a compilation and evaluation of the inputs, outputs, and potential environmental impacts of a product system throughout its life cycle [6]. LCA studies always consist of four main phases, which are covered through ISO standards (DIN 14044; ISO 14040:2006). The first step of the LCA is used to define the goal and scope of the study. The second step is a life cycle inventory (LCI) model through which data is collected and organized. The third step is the life cycle impact assessment (LCIA), used to understand the relevance of all the inputs and outputs in an environmental framework. The fourth step is the interpretation, which is a systematic technique to identify, check, and evaluate information resulting from the LCIA (see Figure 2).

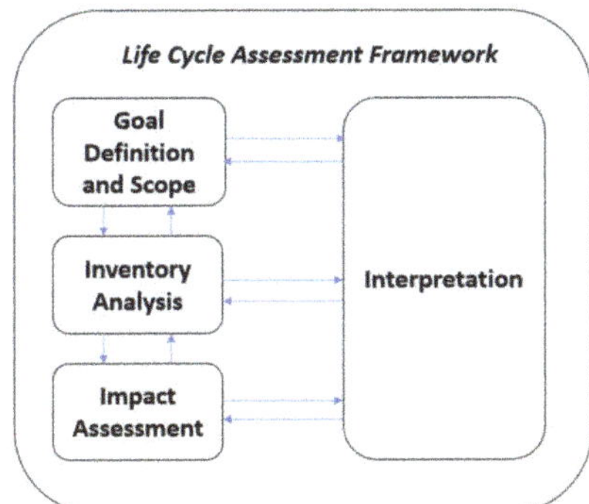

Figure 2. Life cycle assessment (LCA) framework (DIN 14044; ISO 14040:2006).

The environmental impact assessment for this study is completed at the mid-point level. Midpoints are considered to be connections in the cause–effect chain of different impact categories, also known as the problem-oriented approach or classical impact assessment method. Greenhouse gas emissions ($kgCO_2$-eq) were used as an indicator of climate change contribution. The 100-year global warming potentials based on the latest IPCC 2013 were assumed, according to their radiative, forcing capacity relative to the reference substance CO_2. Global warming potential (GWP) during the life cycle stages of a PV system was estimated as an equivalent of CO_2 containing all the significant emissions CO_2 (GWP = 1), CH4 (GWP = 25), N2O (GWP = 298) and chlorofluorocarbons (GWP = 4750–14,400). The calculations were performed using LCA software GaBi with Ecoinvent v2.2+ as back-ground database. GaBi is a process-oriented software, examining the material and energy flows of each step of the production chain. The applied methodology of this study can be divided into four sections which will be briefly described below.

(A) VIPV Use Case Parameters

In the first section, the system boundaries and the input parameters for the operation in urban delivery were clarified and collected for the Use Case StreetScooter. Promising VIPV configuration was defined as the result.

(B) LCI of VIPV—Manufacturing

The second section was dedicated to compiling an inventory of energy and material inputs and outputs over the life cycle of VIPV. Life-Cycle Inventory was completed based on the data from literature. Based on the Inventory, a GaBi model was developed.

(C) LCI of VIPV—Operation and VIPV Energy flow model

The third part included the simulation of VIPV contribution to charging. To simulate the reduction of grid power demand, an energy flow model for the identified reference case was developed.

(D) Evaluation of environmental impacts

In the last part, potential environmental impacts (GWP) related to identified inputs and releases were evaluated. LCA results were compared to grid charging by means of their environmental impacts. The characterization factors were based on IPCC (2013) and should be incorporated for the impact category of global warming potential, which is tracked in kg CO_2 eq.

2.2. Functional Unit, Goal, and Scope

The use case is the light utility battery electric vehicle Work L of StreetScooter. The functional unit for this study is 1 kWh of electricity supplied by the PV system to the battery of the StreetScooter. In comparison to the functional unit of 1 km driven, the emissions of 1 kWh can be calculated more accurately. Furthermore, the chosen functional unit of 1 kWh allows for a direct comparison of effects of charging by PV modules to those due to charging by the grid. Thus, the emissions of VIPV and grid charged BEV can be evaluated more precisely referring to the same functional unit. The operation of the electrical vehicle is set in Cologne, Germany and starts in 2017. Within the scope of this project, the environmental impacts of VIPV are to be studied according to the standard of life cycle assessment ISO 14040:2006. The PV system configuration is based on the first generation of the VIPV panels for the STREET Project with heterojunction silicon PV modules manufactured in China. The analyzed VIPV configuration includes three panels and three control units including the cables mounted on the vehicle roof. The overall capacity of the VIPV system is 930 Wp. The system of the VIPV electricity includes raw material extraction, wafers, crystalline silicon-based heterojunction solar cells and module manufacturing, mounting structures manufacturing, inverters manufacturing, system installation, and the operation.

2.3. Input Parameters for the Life Cycle Inventory

The production process of a typical commercial crystalline silicon solar cell is modelled based on the existing datasets describing the supply chain [7] (see Figure 3). Input parameters of the manufacturing of the PV control unit (PVCU) as well as the vehicle integration process were added based on internal communication in the project STREET. The electricity consumption on all process levels is modelled following specific electricity mixes corresponding to China (CN) or Germany (DE), respectively, based on the Ecoinvent datasets.

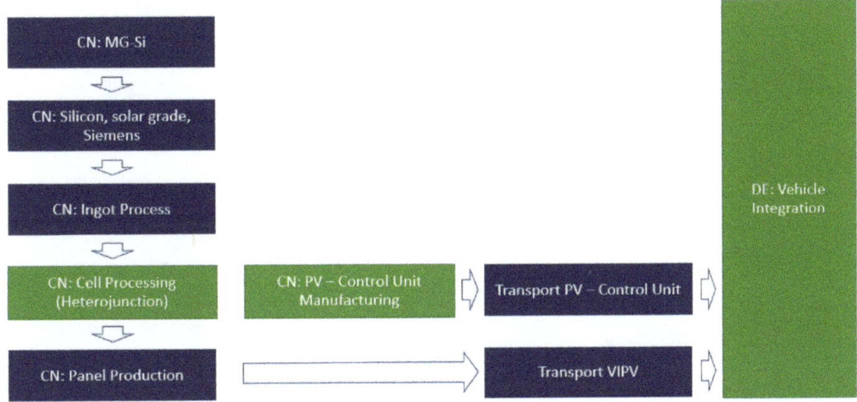

Figure 3. Vehicle integrated photovoltaics (VIPV) system value chain: process flow diagram. CN means China, DE Germany.

All input parameters of the manufacturing and vehicle integration process are described in Table 1. The exact location of manufacturing plants is undocumented and unknown. However, it can be assumed that the location of these plants is somewhere within China. Modelling of the transportation was based on the standard distances as suggested in the Guideline for PC LCA [7]. Metal parts were commonly reported with 200 km train and 100 km truck transportation in China. Additionally, transoceanic transport from China to Belgium was estimated to be 19,994 km based on searates.com data. In Europe, lorry transport from Antwerp (Belgium) to Cologne (Germany), a total of 500 km, was used.

Table 1. Input parameters of the manufacturing process and vehicle integration Ppocess.

Parameter		Based on	Comment
Wafer	Type: n-type c-Si	(b)	Wafer (Solar Grade)
	Thickness: 180 nm	(a)	Wafer thickness
Cell	Technology SHJ c-Si		SHJ cell processing adopted due to STREET requirements, μc-SiOx:H
	Area: 239 mm^2	(a)	156 × 156 mm^2
	Efficiency: 22.5%	(c)	Describes the efficiency of the solar cells
Panel	Efficiency: 19.7%	(c)	Describes the efficiency of the solar cells
	Glass thickness: 2 mm	(c)	The thickness of the solar glass used on the front side of the solar cell
	Back: EVA Back Foil	(a)	EVA back foil configurations based on the Guideline
	Cell number per panel: 72	(b)	Standard-based on the Guideline
	Panel Number: 3		
Vehicle Integration	Total PV Area: 4.8 m^2	(b)	
	PVCU Number 3		
	Mounting: on the rooftop		
	Integration with Bosch Profiles		

(a) Guideline (Frischknecht et al. 2015) [7]; (b) STREET Internal Expert Judgement; (c) LCI on SHJ Cells (Louwen et al. 2016; Olson et al. 2013) [8,9].

For the solar cells in the VIPV, heterojunction technology (SHJ) was chosen due to the best trade-off between efficiency and costs. Thus, in the LCA, the cell heterojunction process was described by the following process steps taken from [8,10] and shown in Table 2. The metallization of the front side requires a double print of the standard amount of silver paste and sputtered aluminum closed back side. The LCI data on material and energy consumption were added for heterojunction cell processing, referring to [8–10].

Table 2. Process steps of heterojunction cell.

Process Step	Material	Description
Metallization front	Ag print	Screenprint
TCO	ITO	Sputtering of indium-tin-oxide
Emitter	a-Si: H (p)	ALD—atomic layer deposition
Passivation	a-Si:H (i)	Deposited by PECVD
BSF	a-Si:H (n)	Back surface field
Metallization back	Ag print	Screenprint

2.4. Input Parameters for the Energy Flow Model

The main factor for the estimation of PV electricity generation is the effective solar irradiance, which depends on the route and location, season, time, and module configuration and orientation. For the reference case of the LCA, the location for the operation was set in Cologne, Germany. The hourly global horizontal solar irradiance was defined by averaging hourly incident global horizontal radiation data extracted from the PVGIS database. The on-board generation of electricity was simulated based on degradation, system losses, and shadowing factor (see Table 3). A 19.7% module efficiency was assumed [9]. In line with IEA-PVPS methodology guidelines [7], degradation of 0.7% per year was applied. Operation time of the reference case was set to 8 years, based on data of LUVs in delivery services [11].

Table 3. Input parameters for the operation of the VIPV.

Parameter	Value	Unit
Capacity	930	Wp
Efficiency	19.7	%
Degradation	0.7	%
Operation lifetime	8	a
Location	Cologne (50.938, 6.954)	Lat/Lon
Database	PVGIS-CMSAF	/

According to the literature guidelines, efficiency for the VIPV system was estimated. Due to dynamic shading, an average 70% performance compared to residential PV was assumed [3]. Furthermore, generated energy cannot be used directly for traction of the vehicle and must be stored in the battery, where DC-Charging/discharging loss of 2% appears. Additional loss of 5% was considered due to the DC/DC converter. The loss of the MPP tracking additionally limits its efficiency in the model to 95% [3]. A performance loss of 9% due to temperature increase and low irradiance was assumed [5]. The overall average efficiency losses of the VIPV system is to be found in Table 4.

Table 4. VIPV system efficiency.

Loss Coefficient	Changes in Output (%)
MPPT loss	−5
Temperature/low irradiance	−9
DC/DC conversions	−5
DC charging/discharging loss	−2
Average shadowing factor	−30

2.5. Input Parameters of the Grid Charge

The grid mix in the location of the charge was analyzed regarding its carbon intensity. The emissions of the grid can vary massively depending on the different power plants. Fossil power plants dominate the power generation in Germany. Acknowledged studies usually consider annual average carbon footprints of the grid power plants caused by the life cycle (construction, fuel production, operation, etc.) [12]. Hourly average emissions of the German electricity mix vary depending on the day and night times. The German electricity mix was modelled using SMARD electricity generation data from 2017 and utilized for the projection of the future scenario [13].

The reference scenario follows the pathway of technological development as far as possible, according to the goals set by politics. The target of the electricity sector in Germany for 2030 is 180–186 Mio t. Until 2028, the annual electricity mix GWP is expected to decrease by 2% per year [1]. Table 5 gives an overview of the emissions of different electricity sources, found in [14].

Table 5. Emission factors of electricity sources.

Electricity Source	(g CO_2 eq./kWh)	Reference
Biomass	272	[15]
Hydropower	3	
Pumped hydro	26	
Wind offshore	6	
Wind onshore	11	
Photovoltaics	67	[12]
Geothermal	192	
Lignite	1142	
Coal	815	
Natural gas	374	
Nuclear	32	[14]
German mix average	486	[12]

2.6. Reliability of the Data

The LCI in this study was based on extracting the data from reliable literature. The commercial LCA software GaBi Version 8.7.1.30 was used to model and calculate the LCI and impact assessment results. Essential materials, electricity mixes were calculated based on data represented by the Ecoinvent database unless otherwise noted. The International Energy Agency (IEA) developed guidelines to make the LCAs of PV systems more consistent and to enhance quality and reliability. Data on production is mainly based on these guidelines and LCIs of photovoltaics [7], additionally considering the heterojunction process of [9,10]. Some values for the Vehicle Integration Process and PVCU were adjusted after internal communication in STREET. The reason for adjustment was mainly a lack of access to the supply chain model data. The data used for this LCA varies in quality and reliability. To limit the resulting uncertainty, the differences of the data sources were analyzed and scored referring to the Quality Pedigree Matrix Flow Indicators determined by DIN 14044. Due to the above-mentioned conditions, the scores for each step were evaluated in Table 6. The highest score shows the lowest uncertainty and data scored with 5 shows the highest uncertainty.

Table 6. LCA data quality.

Process	Data Source	Quality	Comment	Flow Score
Feedstock, ingot, wafer production	(Frischknecht et al. 2015) Material flow: Ecoinvent database	Primary data, measured	High variability of process data, low uncertainty, verified data based on measurements with less than 6 years of difference	2
Wafer cleaning, texturing, PECVD of a-Si layers, TCO deposition, contacting, wiring	(Louwen et al. 2016; Louwen et al. 2012b; Olson et al. 2013) Material flow: Ecoinvent database	Primary data, measured, Ecoinvent processes updated based on updated data	Low variability of process data, low uncertainty, based on measurements with less than 6 years of difference	1
PVCU	Expert judgement—continental material flow: Ecoinvent database	Primary data, adjusted and verified by the STREET experts	Low variability of process data, higher uncertainty, verified by STREET Experts, less than 6 years of difference	2
Module assembly	(Frischknecht et al. 2015) Material flow: Ecoinvent database	Primary data, measured	Low variability, higher uncertainty, verified precise data based on measurements with less than 6 years of difference	2
Vehicle integration	Expert judgment: StreetScooter Material flow: Ecoinvent database	Estimated data based on metadata, verified by the STREET concept	High variability, high uncertainty, documented estimate, verified by STREET Experts, less than 6 years of difference	3
Operation	(PV-Powered Vehicle Strategy Committee 2019) Material flow: Ecoinvent database	Estimated data based on metadata, lacking measurements of PV output and maintenance	High uncertainty, high variability, documented estimate, verified by STREET Experts, less than 6 years of difference	3

3. Results

This section presents the results of the LCA study completed to the mid-point level.

3.1. Manufacturing Process of the VIPV

The results of the analysis of the manufacturing phase [$kg\,CO_2eq$] demonstrate the impact before the operation starts. The manufacturing process VIPV shows similar results to other PV systems. The most dominant contributor to this phase is the Solar-Grade Process. It is responsible for 444.30 kg CO_2 eq, a third of total emissions. The process of integration of the cells into the panel emits 235.24 kg CO_2 eq. The calculated total amount of emissions during the manufacturing process is 1143.12 kg CO_2 eq (see Figure 4).

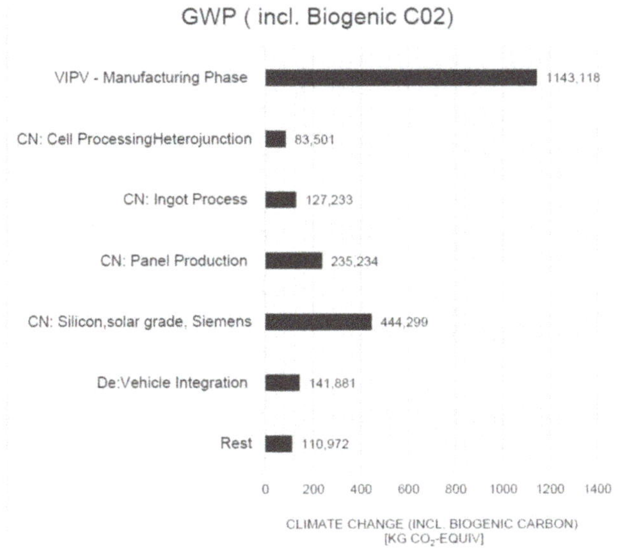

Figure 4. Results of the LCA, the manufacturing phase global warming potential (GWP) = 1143 [kg CO$_2$ eq].

3.2. Operation Phase of the VIPV

The on-board generation of electricity was simulated based on the assumptions on degradation, system losses, and shadowing factor, as previously described in Section 2.4. While driving the EV, the batteries will discharge and will recharge again using the on-board PV modules. The degree of VIPV's impact was expected to vary with the usage patterns: different daily driving distances have different depths of discharge corresponding to daily driving durations. In this study, all incoming irradiance during the day is used, assuming energy is being collected energy and the battery is being charged, even if not driving. The results of the energy flow model are shown in Table 7.

Table 7. Results of the energy flow model.

Parameter	Value	Unit
Average yield in urban area	936	(kWh/kWp)
Average annual VIPV electricity production on board	479	(kWh/year)
Total production VIPV	3738	(kWh)

For the reference scenario of 8 years operation and a shadowing factor of 30%, the VIPV contribution is 3738.116 kWh. Prolonged operation of 12 years generates 5526.702 kWh in total.

3.3. Comparison to the Emissions of the Grid Charge

For the same amount of energy, if the grid would be used, 1630 kg CO$_2$-eq for 8 years and 2267 kgCO$_2$-eq for 12 years were calculated. The losses appearing due to grid distribution were not calculated, because the emission factor is already based on an energy consumption perspective.

Main findings of the comparison with grid electricity show: VIPV can improve the carbon footprint for the reference case of an average shadowing factor of 30% and 8 years of operation time. For the functional unit of 1 kWh of on-board generated PV electricity, the emission factor of 0.357 kgCO$_2$-eq/kWh is calculated for the reference case. In comparison, the average grid emissions for the operation time are expected to be 0.435 kgCO$_2$-eq/kWh.

Considering the data quality of the LCA, reduction of emissions of the functional unit for the reference case compared to the grid is about 18%. The holistic view of the results for the reference case shows 3738 kWh VIPV contribution. For the functional unit of 1 kWh of on-board generated PV electricity, the emission factor of 0.357 $kgCO_2eq/kWh$ is calculated. In comparison, the average grid emissions for the operation time are expected to be 0.435 $kgCO_2eq/kWh$. Compared to the estimated grid average, about 18% less emissions per kWh are caused by VIPV. Projected contribution of VIPV was replaced by grid charging to find out in which operation year VIPV have fewer emissions than the grid and thus calculate the "ecological break-even point". In the previously described reference case, this point is achieved in the year 2022. That means that after 6.5 years of operation, the ecological impact of VIPV equals the impact of the grid charge. However, an increasing shadowing factor of mobile application causes a significant growth of emissions per kWh.

3.4. Sensitivity Analysis

The results of the study are wide-ranging. The variations mainly arise from system operating assumptions (e.g., solar irradiation, system lifetime, shadowing factors) and technology improvements (e.g., electricity consumption for manufacturing processes). In this section some adjustments of the reference case (8 years of operation, 0.7% degradation, and 30% shadowing factor) are considered.

PV-generated power is an essential variable for the reduction in emissions. By increasing the shadowing factor, emissions per kWh grow significantly. An emission factor of 0.357 $kgCO_2$-eq/kWh is calculated for the reference case. The increased shadowing factor of 40% results in 0.435 $kgCO_2$-eq/kWh, which equals the average emissions of the future grid electricity. As shown in Figure 5, the ecological benefit over the grid charge disappears completely when the shadowing factor reaches 40%.

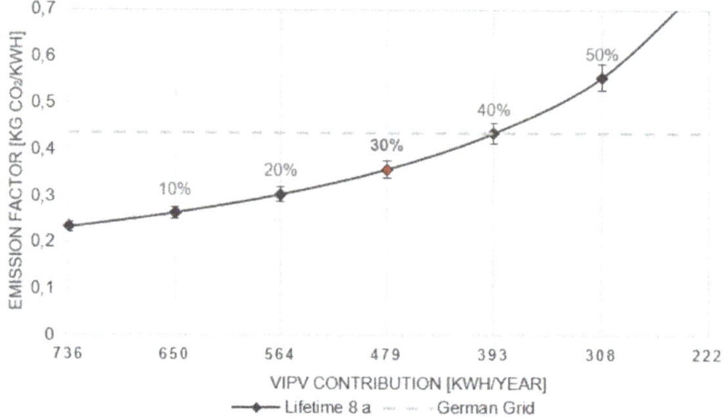

Figure 5. Emissions depending on the shadowing factor.

Sensitivities show that if the VIPV is used for a prolonged life of 12 years, the emission factor of the produced electricity decreases to 0.221 $kgCO_2$-eq/kWh. A reduction of 38% (0.136 $kgCO_2$-eq/kWh) compared to the reference case of 8 years is noted. The average grid mix emissions of prolonged use decrease to 0.409 $kgCO_2$-eq/kWh. Comparable results can be achieved with a shadow factor of 55% or an average annual VIPV generation of about 260 kWh/a. Lifetime extension of the vehicle operation will automatically result in a reduction of the emissions per produced kWh of the VIPV. Figure 6 demonstrates the potential of the longer operation phase for different shadowing factors.

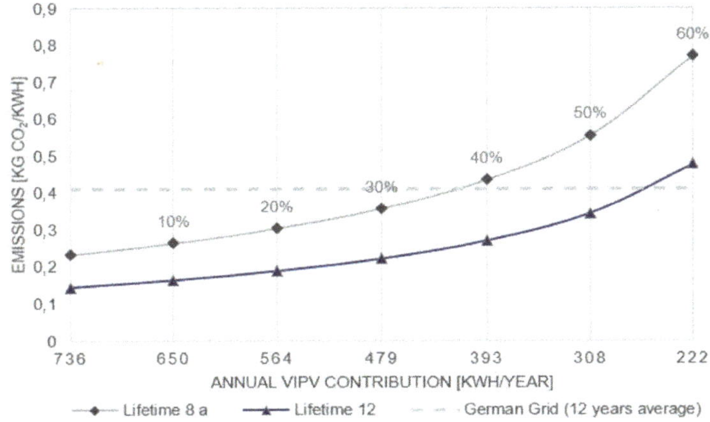

Figure 6. Emissions of "prolonged use" scenario.

Based on findings of the sensitivity analyses, the highest potential for emission reduction can be confirmed for a "green" electricity scenario, where renewable electricity is used for the manufacturing process. The emission factor of 0.831 kgCO$_2$/kWh for the electricity mix of China used for the simulation of the reference case is based on the GaBi Education Database from 2017. With the increasing share of renewable energy in the electricity mix, lower GWP impact will arise from the production phase of the VIPV. Using green electricity has the potential to be almost carbon-free, as is the case for today's hydropower. For "green" electricity, assumptions of hydro plants with average emissions of 0.003 kgCO$_2$-eq/kWh were used to cover the energy need of the manufacturing phase in China [15]. As illustrated in Figure 7, the emissions decrease from 0.357 to 0.230 kgCO$_2$-eq/kWh for the shadowing factor of 30%.

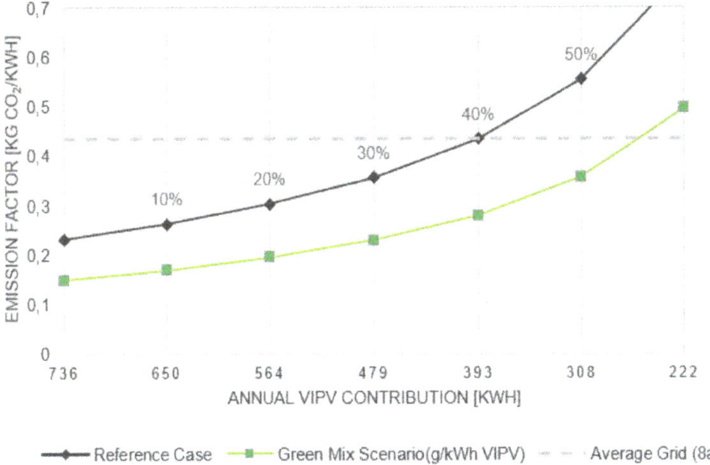

Figure 7. Emissions of "green manufacturing" scenario.

4. Discussion and Conclusions

The study reports the unique observation that placing a PV system on-board of an existing StreetScooter can improve the carbon footprint of the generated electricity for the reference case of an average shadowing factor of 30% and 8 years of operation time. The ecological benefits of PV-powered light utility vehicles are confirmed for the reference case of the StreetScooter. Yet, the results of the LCA show that viability is heavily dependent on the vehicle's deployment region and usage scenario. Main findings of the comparison to the grid electricity show: VIPV can improve the carbon footprint for the reference case of an average shadowing factor of 30% and 8 years of operation time. For the functional unit of 1 kWh of on-board generated PV electricity, the emission factor of 0.357 $kgCO_2$-eq/kWh is calculated. In comparison, the average grid emissions for the operation time are expected to be 0.435 $kgCO_2$-eq/kWh. Considering the data quality of the LCA, reduction of emissions of the functional unit for the reference case compared to the grid is about 18%. By increasing the shadowing factor, emissions per kWh grow significantly. The ecological benefit to the grid charge disappears completely when the shadowing factor reaches 40%. However, if the operation time is prolonged to 12 years, the shadowing factor can reach 55%, having similar emissions to grid charge. For the reference case with 30% shadowing, a reduction of 38% compared to 8 years in use can be noted. For this case, 0.221 $kgCO_2$-eq/kWh is estimated for the functional unit.

One of the key challenges of this work was finding an appropriate vehicle usage model to reproduce the ratio of using solar power and performance assessment of Maximum Power Point Tracker (MPPT) algorithms for VIPV. Tests with radiation sensors investigating shading and reflection conditions are suggested. Numeric simulation of VIPV output test drives with irradiance profiles of routes should include different vehicle usage times, effects of panel position and movement. Additionally, it is necessary to address the electrical and technical issues.

For the recycling process, no established and reliable routes were found. As to the knowledge of the author, no study provides details on the LCI with the input and output of every process stage. However, if material depletion is considered, recycling is crucial, and further research should include recyclability options. Since second use is an important issue, the mounting structure, removable, and lightweight, must become a priority for research. Furthermore, a scenario of VIPV connection to the public grid while parking during weekends, in which the surplus of unused electricity can be fed into the grid, seems to be realistic. Vehicle2Grid (V2G) concepts can be very profitable, but first, the dependence on the state of charge (SOC) of the battery including an ageing model of the battery with charging and discharging losses should be analyzed. Enhanced communication and cooperation between automotive companies and PV players can contribute to the positive image of vehicle integrated photovoltaic systems in order to achieve the goal to change the image of VIPV. Likewise, international methods for evaluating the reduction of emissions of PV-powered vehicles can help to communicate the created value for the different driving and charging behaviors. To contribute to the growth of the VIPV market, governments willing to achieve emission goals must support the standardization of the technology. To solve this problem, international methods of evaluating added value on the reduction of grid power and ecological benefits are required.

Author Contributions: Conceptualization O.K., K.D.; methodology O.K., J.M.; software, validation, investigation, resources, data curation and writing—original draft preparation O.K.; writing—review and editing A.R., J.M., K.D.; visualization, O.K.; supervision J.M., A.R., project administration K.D. All authors have read and agreed to the published version of the manuscript.

Funding: This work was supported by the Federal Ministry for Economic Affairs and Energy in the framework of the STREET project (grant: DB001618).

Acknowledgments: This work is based on the Master's thesis by Olga Kanz at the Hochschule Köln 2019. The authors appreciate the support of the research group of IEA Photovoltaic Power System Programme Task 17, which focuses on possible contributions of photovoltaic technologies to transport and support from Toshio Hirota and Keichi Komoto. Costs for publishing are covered by Forschungszentrum Jülich GmbH.

Conflicts of Interest: The authors declare no conflict of interest. The founding sponsors had no role in the design of the study; in the collection, analyses, or interpretation of data; in the writing of the manuscript, and in the decision to publish the results.

References

1.	European Environment Agency (EEA). Progress of EU Transport Sector Towards Its Environment and Climate Objectives. 2018. Available online: https://www.eea.europa.eu/themes/transport/term/term-briefing-2018 (accessed on 30 September 2020).
2.	Helmers, E.; Dietz, J.; Hartard, S. Electric car life cycle assessment based on real-world mileage and the electric conversion scenario. *Int. J. Life Cycle Assess.* **2015**, *22*, 15–30. [CrossRef]
3.	New Energy and Industrial Technology Development Organization. NEDO, PV-Powered Vehicle Strategy Committee. 2018. Available online: https://www.nedo.go.jp/content/100885778.pdf (accessed on 30 September 2020).
4.	Rizzo, G. Automotive application of solar energy. In Proceedings of the 6th IFAC Symposium Advances in Automotive Control, Munich, Germany, 12–14 July 2010.
5.	Araki, K.; Ji, L.; Kelly, G.; Yamaguchi, M. To Do List for Research and Development and International Standardization to Achieve the Goal of Running a Majority of Electric Vehicles on Solar Energy. *Coatings* **2018**, *8*, 251. [CrossRef]
6.	International Organization for Standardization. *Environmental Management—Life Cycle Assessment—Requirements and Guidelines (ISO 14044:2006 + Amd 1:2017)*; German Version EN ISO 14044:2006; ISO: Geneva, Switzerland, 2006.
7.	Frischknecht, R.; Itten, R.; Sinha, P.; de Wild-Scholten, M.; Zhang, J. *Life Cycle Inventories and Life Cycle Assessments of Photovoltaic Systems, IEA PVPS Task 12, Subtask 2.0*; Life Circle Assessment Report IEA-PVPS 12; International Energy Agency: Paris, France, 2015.
8.	Louwen, A.; Van Sark, W.; Schropp, R.E.I.; Faaij, A. A cost roadmap for silicon heterojunction solar cells. *Sol. Energy Mater. Sol. Cells* **2016**, *147*, 295–314. [CrossRef]
9.	Olson, C.; de Wild-Scholten, M.; Scherff, M. Life Cycle Assessment of Hetero-junction Solar Cells. In Proceedings of the 26th European Photovoltaic Solar Energy Conference and Exhibition, Hamburg, Germany, 5–6 September 2011.
10.	Louwen, A.; van Sark, W.; Turkenburg, W.C.; Schropp, R.; Faaij, A.C. R&D Integrated Life Cycle Assessment: A Case Study on the R&D of Silicon Heterojunction (SHJ) Solar Cell Based PV Systems. In Proceedings of the 27th European Photovoltaic Solar Energy Conference and Exhibition, Frankfurt, Germany, 23–27 September 2012; pp. 4673–4678.
11.	Hacker, F.; Waldenfels, R.; Moschall, M. *Wirtschaftlichkeit von Elektromobilität in Gewerblichen Anwendungen. Betrachtung von Gesamtnutzungskosten, Ökonomischen Potenzialen und Möglicher CO_2EQ—Minderung im Auftrag der Begleitforschung zum BMWi Förderschwerpunkt IKT für Elektromobilität*; Öko-Institut: Berlin, Germany, 2015.
12.	Icha, P. Entwicklung der Spezifischen Kohlendioxid-Emissionen des Deutschen Strommix in den Jahren 1990–2018. 2019. Available online: https://www.umweltbundesamt.de/sites/default/files/medien/1410/publikationen/2019-04-10_cc_10-2019_strommix_2019.pdf (accessed on 30 September 2020).
13.	Bundesnetzagentur für Elektrizität, Gas, Telekommunikation, Post und Eisenbahnen. SMARD. 2020. Available online: https://www.smard.de/home (accessed on 22 June 2020).
14.	May, J.F.; Kanz, O.; Schürheck, P.; Fuge, N.; Waffenschmidt, E. Influence of usage patterns on ecoefficiency of bat-tery storage systems for electromobility and home storage. In Proceedings of the 3rd Product Lifetimes and the Environment Conference, Berlin, Germany, 18–20 September 2019.
15.	Memmler, M.; Lauf, T.; Schneider, S.; Dessau-Roßlau, U. Emissionsbilanz Erneuerbarer Energieträger—Bestimmung der Vermiedenen Emissionen im Jahr 2017. 2018. Available online: https://www.umweltbundesamt.de/sites/default/files/medien/1410/publikationen/2018-10-22_climate-change_23-2018_emissionsbilanz_erneuerbarer_energien_2017_fin.pdf (accessed on 30 September 2020).

Article

Sustainable E-Bike Charging Station That Enables AC, DC and Wireless Charging from Solar Energy [†]

Gautham Ram Chandra Mouli *, Peter Van Duijsen, Francesca Grazian, Ajay Jamodkar, Pavol Bauer * and Olindo Isabella

Department of Electrical Sustainable Energy, Delft University of Technology, 2628 CD Delft, The Netherlands; P.J.vanDuijsen@tudelft.nl (P.V.D.); F.Grazian@tudelft.nl (F.G.); ajay.j001@gmail.com (A.J.); O.Isabella@tudelft.nl (O.I.)
* Correspondence: G.R.Chandramouli@tudelft.nl (G.R.C.M.); p.bauer@tudelft.nl (P.B.)
† This paper is an extended version of our paper published in 2018 in the European Conference on Power Electronics and Applications (EPE-ECCE Europe), Riga, Latvia, 20 September 2018.

Received: 24 May 2020; Accepted: 4 July 2020; Published: 10 July 2020

Abstract: If electric vehicles have to be truly sustainable, it is essential to charge them from sustainable sources of electricity, such as solar or wind energy. In this paper, the design of solar powered e-bike charging station that provides AC, DC and wireless charging of e-bikes is investigated. The charging station has integrated battery storage that enables for both grid-connected and off-grid operation. The DC charging uses the DC power from the photovoltaic panels directly for charging the e-bike battery without the use of an AC charging adapter. For the wireless charging, the e-bike can be charged through inductive power transfer via the bike kickstand (receiver) and a specially designed tile (transmitter) at the charging station, which provides maximum convenience to the user.

Keywords: battery charger; electric bike; electric vehicle; photovoltaic system; power converter; wireless power transfer

---

1. Introduction

Electric vehicles (EVs) have several advantages over conventional fossil fuel-powered vehicles such as zero tailpipe emissions, higher tank-to-wheel efficiency, low noise, and full torque at zero speed. With the increased use of electric vehicles in the form of e-bikes, electric cars, and electric buses, there is increased emphasis to make sure that the electricity used to charge the EVs are sustainable as well [1,2]. As shown in Figure 1, the average emission from electricity production for various European countries shows a wide variation of 13–819 g of CO_2 equivalent per kWh based on the electricity generation mix in 2016 [3]. The European Union wide average is 296 g/kWh, which is primarily driven by the 43% share of fossil fuels in the gross electricity generation. Assuming a modest energy efficiency of an electric car to be 5 km/kWh, the CO_2 emission of 13–819 g/kWh translates to a 2.6–163 g/km. Therefore, to make EVs truly sustainable, it is essential to charge EVs from sustainable sources of electricity, such as wind and solar energy [4–7].

Using solar energy for charging EVs is attractive due to the possibility of distributed photovoltaic (PV) generation in locations close to where EVs are parked. Specifically, charging electric vehicles from PV panels at workplaces has significant potential for the future due to several reasons.

- There is excellent synergy between the hours of sunshine and the working hours at workplaces.
- The solar panels can be installed on the large roof area of office buildings, factories, or parking lot. This potential is largely untapped today.
- Reduced peak power and energy demand on the grid as the EV charging power is locally produced from PV [6].

- Reduced cost of EV charging due to lower levelized cost of solar energy when compared to grid electricity tariffs [8,9].
- The EV can be used as an energy storage buffer for PV, and this could be very useful when PV feed-in tariffs are gradually reduced in the future [9,10].

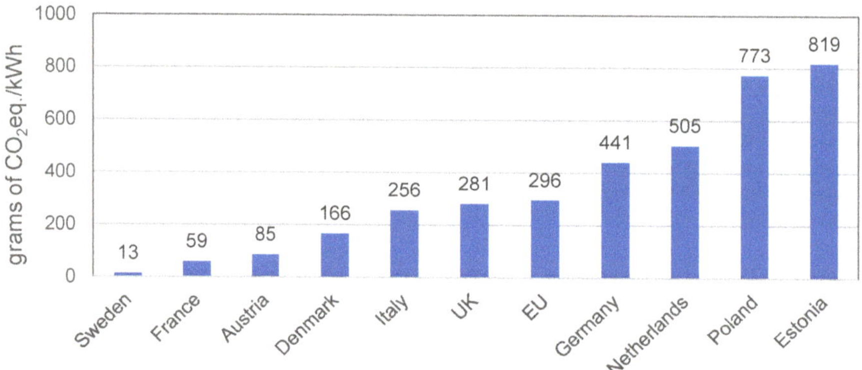

Figure 1. CO$_2$ emission from electricity production for selected European countries for 2016, in g CO$_2$ equivalent/kWh.

Concurrently, the disadvantage of charging EVs at the workplace using solar energy is the low solar generation in winter months. Secondly, there is a lack of charging demand on the weekends for workplaces that have a weekday working week.

1.1. Electric Bikes and Electric Scooters

Electric bikes and electric scooters provide a convenient means of intra-city commute with a multitude of benefits such as door-to-door connectivity, low (indirect) emissions, reduced traffic, and parking congestion and a fraction of the energy usage of an electric car [11]. In 2016, around 30% of the 928,000 bikes sold in the Netherlands were e-bikes [12]. An e-bike can travel up to 25 km/h using a motor of up to 250 W (up to 1000 W for high-speed e-bikes) and uses a 12 V–48 V battery with an energy capacity of 0.2–1 kWh. Electric mopeds (including speed pedelec), on the other hand, can go up to 45 km/h using a motor of 1–4 kW and typically uses a 48 V battery of 1–5 kWh. Furthermore, e-bikes have an extremely low energy consumption in the range of 5–15 Wh/km depending on the drivetrain efficiency, riding behaviour, tire characteristics, and the combined weight of the bike and rider. This is much lower than the 150–200 Wh/km energy consumption of an electric car.

By providing a charging facility at the workplaces and powering it with solar energy, e-bikes can be made into a fully sustainable means of daily commute [13,14]. The focus of this paper is on the development of a charging station to sustainably charge e-bikes at the workplace using solar energy. The charge station is shown in Figure 2 and provides three modes of charging: AC, DC, and wireless charging, respectively.

1.2. State-of-the-Art

The most common design for a solar e-bike charging station is to use the AC low voltage grid for power exchange between the PV and the EV. A solar inverter with maximum power point tracking (MPPT) feeds the solar power to the AC grid [15–19]. A standard e-bike AC power adapter is then used for e-bike charging. Even though there have been recent studies to simplify the complexity of the e-bike power adapter [20,21], the disadvantage of such designs is due to the unnecessary power conversion from DC to AC and back, even though both the solar panels and e-bike battery operate on DC [18,22,23]. DC charging the e-bikes directly from the PV would reduce the power conversion stages, and it would not require the cyclists to bring the power adapter but only requires a DC cable. However, e-bike DC charging is still manufacturer-specific, and, therefore, no standard exists except for the consensus on typical battery voltage levels of 24 V, 36 V, and 48 V. The challenge is primarily due to lack of standard connectors and communication protocols for charging control and safety that exist between the power adapter and the bike battery, which varies across manufacturers.

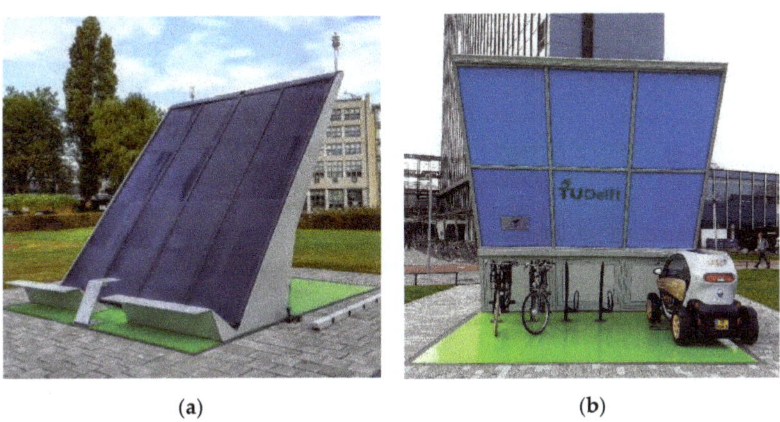

(a)   (b)

Figure 2. (**a**) Front view of solar e-bike charging station showing solar panels. (**b**) Back view showing the charging status display screen, the e-bikes, and Twizy EV charging.

The next step for a more user-friendly and safer experience in e-bike charging would be the transition from plug-in charging methods to wireless charging. This solution was already proposed in Reference [24], where a 100 W e-bike wireless charger has been demonstrated to have an efficiency above 90%. A considerable number of e-bike wireless charging systems use flat-air coils to realize the power transfer [24–28] because they are lightweight. However, air coils might produce a magnetic field higher than the safety limit for the general public in their proximity [29]. Other solutions use a ferromagnetic core to improve the coupling between the coils [30–34], but, at the same time, they introduce extra elements on the e-bike, which do not have any other purposes. In Reference [35], a review of solar-powered wireless charging systems for light electric vehicles is presented and shows 400 W–5 kW designs operating at a resonant frequency in the range of 8.7–100 kHz with a 2.8–20 cm air gap and an efficiency in the range of 75–95%.

Lastly, two key features that distinguish the charging station are the physical design of the structure and the use of energy storage. The usual method to install solar panels is to place them on the parking lot or the rooftop or façade of buildings, while the less common technique is to use a dedicated solar park-port [15–17,36]. The critical design tradeoffs are the cabling losses, construction/installation costs, accessibility, and need for large electrical cabinets to store the electronics. This becomes even more relevant for charging stations with battery storage due to their large size and weight. Most of the existing e-bike charging stations lack an integrated battery storage. This prevents the station from being used in an off-grid mode, especially on days where the solar insolation is very low during the year [18].

1.3. Contributions

This paper presents the development of a novel e-bike charging station that provides sustainable direct DC charging from solar photovoltaic energy, as shown in Figure 2. Using Reference [14] as a starting point, this paper provides the detailed design of the subcomponents of the charging station. The contributions are the following.

- The station offers three methods for charging the e-bikes, namely AC charging, wireless charging, and direct DC charging. The DC charging is implemented in such a way that users no longer need a power adapter and only a DC cable for charging.

- Wireless charging of e-bikes is possible via inductive coils made of ferromagnetic material with the transmitter coil located under the floor tile of the station, and the receiver coil is integrated into the bike kickstand. In this way, the kickstand of the bike is used for both parking, wireless charging, and communication. A novel auto-resonant frequency control is developed for misalignment tolerance of wireless charging.

- The charging station has a bidirectional hybrid inverter and an integrated storage on a DC nano-grid, which facilitates both grid-connected and off-grid operation [37].

- 3D modelling via Sketchup is used for PV shading analysis of the built environment, which provides much higher accuracy of yield estimation than the methods used in Reference [14].

- The physical structure of the charging station and the electrical design are cohesively integrated, which results in an environmentally-integrated PV system (EIPV), which is ergonomic, modular, safe, and designed for outdoor weather conditions. The EIPV provides 3-m^3 internal cabinet space for all electronics, which removes the need for additional cabinets and, hence, increases the overall aesthetics.

1.4. Structure of the Paper

In Section 2, the system design of the e-bike charging station together with the characteristics of the charging demand and battery storage is explained. Section 3 investigates the PV system design and modelling to determine the energy yield, shading conditions, and optimal orientation of the PV system. Sections 4–6 explain the AC, DC, and wireless charging system of the e-bike, including simulations and building of the experimental prototype. Section 7 presents the development of the charging station, experimental measurements of yield, and power management strategies.

2. Solar E-Bike Charging Station

2.1. System Design

The electrical schematic and the specifications of the key components of the e-bike charging station are shown in Figure 3 and Table 1, respectively. The core of the system is a 48 V DC nano-grid, which supports the power exchange between all the components. The PV generation consists of eight modules arranged in four parallel strings. A Victron BlueSolar 150/85 maximum power point tracking (MPPT) converter is used to process power from the PV panels to charge the 48 V battery bank [38]. Isolated DC-DC converters and a high-frequency DC/AC inverter are connected to the 48 V DC nano-grid for DC and wireless charging of e-bikes, respectively.

The connection between the DC nano-grid and the 50 Hz AC grid is realized by a Victron Multiplus 48/3000 hybrid bidirectional inverter [38], which is equipped with two AC outputs. One output is connected to the single-phase AC grid. The second output powers both the e-bike AC charger and the lighting and system monitoring unit, such that the e-bike charging station supports the off-grid operation. Lastly, a Lufft WS503-UMB weather station measures the incoming solar radiation, ambient temperature, and local wind speed for monitoring and research purposes [39]. The system control and monitoring are done using a central Raspberry Pi controller, and a website provides users with feedback and data for scientific research (http://solarpoweredbikes.tudelft.nl).

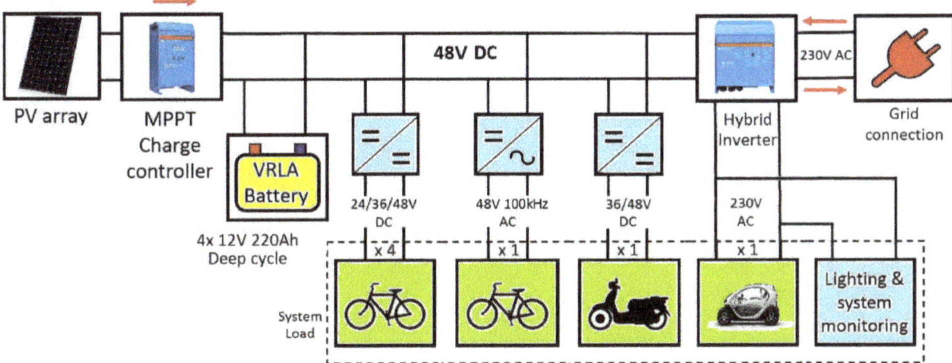

Figure 3. Schematic of the solar e-bike station with 48 V DC interconnection that facilities power exchange between the solar panels, e-bike chargers, and the AC grid.

Table 1. Specifications of the solar e-bike charging station.

Solar panels	8× Sunpower X20-327-BLK, 327 W
Battery	4× Victron Lead Acid Batteries, 220 Ah, 12 V
MPPT converter	Victron BlueSolar 150/85
Grid Inverter	Victron Multiplus 48/3000 Bidirectional
Weather station	Lufft WS503-UMB
Controller	Raspberry Pi
e-bike Charging	1× AC, 4× DC (10–50 V), 1× Wireless
DC charging	100 W, 24–48 V, with isolation
Wireless charging	200 W, 24–48 V, via kickstand

The power balance equation of the system is:

$$P_{PV} + P_{grid} = P_{batt} + P_{load} + P_{loss} \tag{1}$$

where P_{PV} is the PV power, P_{grid} is the power drawn from the grid, P_{batt} is the charging power of the battery, P_{load} is the power consumption of the station including that of the e-bikes and the baseload, and P_{loss} is the total energy conversion losses.

2.2. Charging Demand

The PV system and storage of the charging station is sized to charge up to five e-bikes and a single e-scooter per day throughout the year, even under low winter insolation. Table 2 shows the specifications of the e-bike, e-scooter, and small car considered for the system sizing [40]. Each e-bike and e-scooter battery has a capacity of 396 Wh and 1920 Wh, respectively. A total of five such e-bikes and one e-scooter, assuming that they arrive with fully drained batteries (the worst case), would translate to a load of 3900 Wh. A baseload of 90 W is consumed by the display, controller, lights, and the weather station, which translates to 2160 Wh. If the charging demand and baseload are combined, the net demand is approximately 6.06 kWh per day. This is the same as the load of one Renault Twizy small electric car. If the charging demand is >6 kWh due to a higher number of e-bikes, then they can be charged by using PV power if the daily PV yield is >6 kWh or using the AC grid power via the DC/AC inverter up to 3 kW*24 h = 72 kWh per day.

Table 2. Specifications of the E-bike and E-scooter.

Model	E-Bike	E-Scooter	Small Car
	Batavus Trento E-go	**Novox C-50**	**Renault Twizy**
Battery type	Lithium-ion	Lithium-ion	Lithium-ion
Capacity (Wh)	396	1920	6100
Voltage (V)	36	48	58
Motor power (W)	250	3500	13,000
Driving range (km)	<120	50–110	100 (ECE-15)
Normal charging time (h)	3.5	5	3.5
Energy consumption (Wh/km)	3.3	17.5–38.4	61

Three types of e-bike charging methods are developed for the charging station: AC, DC, and wireless charging. The benefit of the AC charging is that it can be universally used for charging all e-bikes, e-scooters, and light EVs by using a charging adapter. On the other hand, for safety reasons, the DC charging is limited to 100 W but has the benefit that the users can simply use a DC cable between the station and the e-bike battery for charging. This DC charging negates the need for an AC power adapter and, hence, provides convenience to the user while plugging in and preventing any possible theft of the adapter. In the case of the wireless charging, the motive is to do away with the user's need for cables altogether, which, therefore, increases the user convenience further [29,30].

2.3. Local Storage

The battery plays a crucial role in providing the off-grid capability to the charging station. The station has four lead-acid gel batteries of 220 Ah capacity each. The batteries are series-connected to 48 V and provide a usable capacity of 9.5 kWh, when operated at a maximum depth of discharge of 90%. With ~6 kWh demand per day, it can provide close to 1.5 days of autonomy to the system. Lead-acid gel batteries were preferred due to two reasons. First, they have a much longer lifetime than standard lead-acid batteries, and second, the cost is much lower when compared to lithium-ion batteries.

The preferred charging mode is to charge the batteries directly from the solar panels on DC. In the case of insufficient solar generation, the battery can be charged (and also discharged) from the AC grid using the DC-AC grid inverter. This bidirectional option can be useful for different grid support services such as peak shaving, reactive power compensation, energy arbitrage, or emergency backup power.

3. PV System Design

The 2.6 kW PV system is the primary source of power for the e-bike charging station. The PV generation potential is estimated based on solar insolation, wind speed, ambient temperature, panel orientation, and shading due to the surrounding terrain. The meteorological data Cabauw Experimental Site for Atmospheric Research (CESAR) database is used for the PV system modelling, which has a resolution of 1 min [41].

3.1. PV System Modelling

Using the CESAR data, the incident solar irradiance and the cell temperature of the PV panel are estimated for different azimuth and orientation using Equations (2)–(6) [42–47].

$$G_{dir(\beta,Am)} = G_{DNI}\, cos(\theta_i) \tag{2}$$

$$G_{diff(\beta)} = G_{DHI}\, (1 + cos\beta)/2 \tag{3}$$

$$G_{gnd} = G_{GHI}\,\rho\left(1 - \mu^{svf}\right) \tag{4}$$

$$G = G_{dir(\beta,Am)} + G_{diff(\beta)} + G_{alb} \tag{5}$$

$$T_{cell} = \frac{\Phi G + h_c T_{amb} + h_{r,sky} T_{sky} + h_{r,gr} T_{gr}}{h_c + h_{r,sky} + h_{r,gr}} \tag{6}$$

where G_{dir}, G_{diff}, and G_{gnd} are the direct, diffused, and ground irradiance incident on the module with a tilt β and azimuth A_m, G_{DNI}, G_{DNI} and G_{GHI} are the direct normal, diffuse horizontal, and global horizontal irradiance, θ_i is the angle of incidence of the direct irradiance beam on the panel, $\Phi = 0.727$ is the absorptivity, $\rho = 0.2$ is the albedo (measured using albedometer), μ^{svf} is the sky view factor, h_c, is the coefficient for convective heat transfer $h_{r,sky}$, $h_{r,gr}$ are the coefficient for radiative heat transfer to the sky, and to the ground, respectively, and T_{amb}, T_{sky}, T_{gr}, and T_{cell} are the ambient, sky, ground, and PV cell temperature, respectively. In this case, the heat transfer coefficients are estimated using an iterative procedure, where $T_{sky} = 0.0522\,T_{amb}^{2/3}$. Lastly, the output power of the PV modules P_{PV} can be calculated.

$$P_{PV} = P_{STC}\left(\frac{G}{G_{STC}}\right)\{1 - \gamma(T_{cell} - 25)\} \tag{7}$$

where $P_{STC} = 327$ W is the module power at Standard Test Conditions (STC), $G_{STC} = 800$ W/m^2 is the irradiance under STC, $\eta = 20.3\%$ is the efficiency of the modules, and $\gamma = -0.3\%/°C$ is the module temperature coefficient for the Sunpower X20-327-BLK modules.

3.2. PV System Orientation

Based on the meteorological data of 2013, the yield of the PV system is estimated for different fixed orientations. For the whole year, considering a fixed orientation, a tilt of 28° and azimuth facing south was found to result in maximum annual yield [6]. Figure 4 shows the average daily yield of the south-facing PV system for each month, considering various fixed tilt angles between 0° and 90°. It can be clearly seen how the tilt angle has a major influence on the monthly yield, especially in the summer months.

Figure 4 also shows the average daily yield if a two-axis tracking system is used. First, it can be seen that the tracking system has a much higher yield as it tracks both the azimuth and tilt when various tilt angles are considered in which the azimuth is fixed. Second, it can be seen that the increase in yield due to tracking is very small for the winter months as most of the irradiance is diffused irradiance. Due to the higher cost of the tracker and the need for a fixed structure for the bike shelter, tracking systems were not further considered in this study.

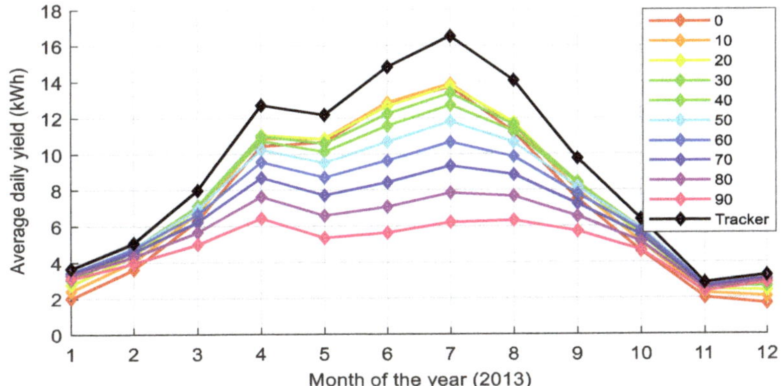

Figure 4. Average daily yield of the south facing 2.6 kW PV system for each month considering various fixed tilt angles compared to the use of a 2-axis tracker.

3.3. Optimal PV System Design

In order to get the maximum annual yield, the PV system could have been oriented at a tilt of 28° and azimuth facing south. However, the motivation is to ensure that there is sufficient generation in December, which has the lowest solar irradiation in the year. For December, the optimal tilt for maximum monthly yield was 65°. This is shown in monthly yield estimation for December in Figure 5a for different PV orientations.

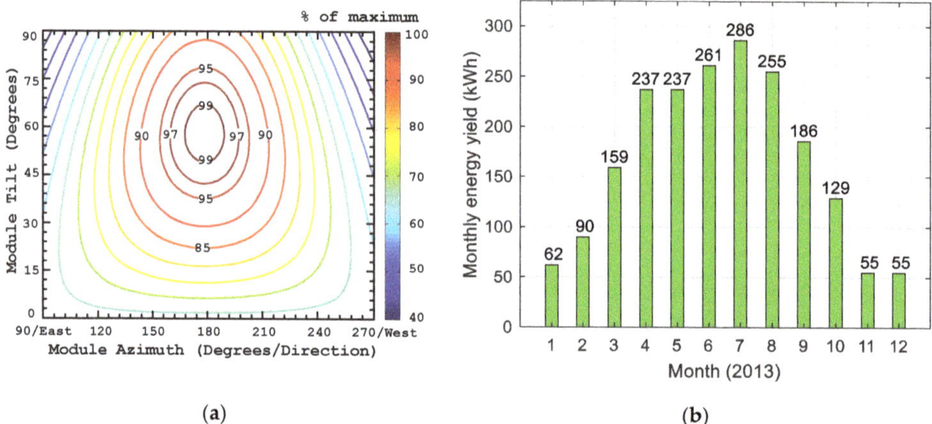

(a)	(b)

Figure 5. (**a**) PV system yield (% of maximum) for different orientations considering December month. (**b**) The estimated monthly energy yield of 2.6 kW PV system tilted at 51° and facing south.

On the other hand, when the tilt is increased from 28° to 65°, the annual yield is dramatically reduced by up to 20%. Hence, a tradeoff was made to set the tilt angle at 51°, which results in <5% reduction in annual yield compared to 28° and 1% reduction in December yield compared to 65°.

3.4. Shading Analysis

Since the e-bike station is installed at the ground level, shading from nearby buildings (especially the tall electrical faculty building) has a significant impact on the PV system output, as shown in Figure 6a. To account for the shading due to the adjacent buildings, their 3D models were made using Sketchup. Figure 6b shows the shading caused by the tall Electrical Faculty building in front of the bike station in the afternoon. The plug-in LSS-Chronolux 3D is then used to estimate the sky view factor $\mu^{svf} = 0.61$ and the per minute shading factor S^{sh} as well as the corresponding direct irradiance including the shading, $G^{sh}_{dir(\beta,Am)}$:

$$G^{sh}_{dir(\beta,Am)} = S^{sh} G_{dir(\beta,Am)} \tag{8}$$

3.5. PV System Yield

Using the 2013 CESAR data and the shading analysis, the PV yield of the system is estimated. The corresponding annual energy yield is 2012 kWh/year, which provides an average daily yield of 5.51 kWh/day, as shown in Figure 5b. Based on the charging demand (Section 2.2), the PV system is sized to a rated power of 2.61 kW$_p$. This means that >90% of the total load demand can be supplied by the PV system on average. At the same time, we have to keep in mind that there is about six times the difference in energy yield between the winter and summer months, as shown in Figure 5b. This results in the load being met on about 180–200 days of the years. That is why a grid connection is required to provide the energy demand in winter, as seen in Equation (1). While the battery cannot provide seasonal storage, it can facilitate off-grid operation and help manage the diurnal solar insolation variation.

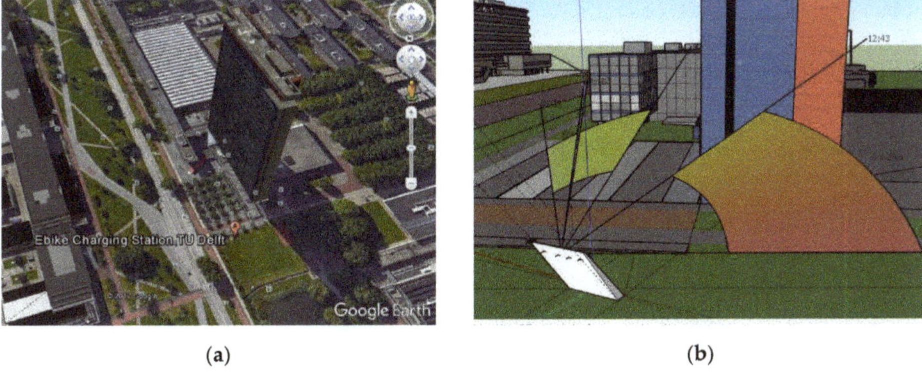

(a)   (b)

Figure 6. (**a**) Google Earth image showing the station location and nearby building. (**b**) Calculation of the shading factor due to nearby buildings using Sketchup.

The highly efficient Sunpower X20-327-BLK modules with a relatively high cost per watt peak are chosen since this reduces the total area occupied by the PV system. The tradeoff is that, in return, the lower area significantly reduces the quantity of steel required for the station structure, which has a relatively higher cost for both material and labour.

4. AC E-Bike Charging

The e-bike charging station provides a single 230 V 50 Hz schuko wall socket for AC charging of e-bikes through the use of a charging adapter. The benefit of the AC charging is that it can be universally used for all light EVs providing up to 16 A charging current, which corresponds to a power of 3.7 kW. This is more than sufficient to charge a small electric car like the Twizy as shown in Figure 2b, which requires a charging current of 10 A. By using the hybrid grid inverter, the AC charging is possible in both grid-connected and an off-grid mode.

5. DC E-Bike Charging

The DC chargers for the e-bikes should provide galvanically-isolated DC power that is controllable in both output voltage level and maximum charging current, based on the e-bike battery. A dual interleaved quasi-resonant flyback converter with digital current-mode control from Involar is chosen for this task [14,48]. As shown in Figure 7a, the primary side of the flyback is operated directly from the 48 V DC nano-grid. The secondary side of the flyback is connected to the DC cable that the user has to plug into the charging inlet of the e-bike battery. The input current ripple is reduced by half due to the interleaving. The magnetically coupled inductors are wound to have minimum parasitic leakage inductance to reduce the Electromagnetic interference (EMI) and the voltage stress on the MOSFET.

5.1. Current Mode Control

Depending on the required output power to charge the e-bike battery, the current level in the primary circuit of each flyback is regulated using the current mode control. Each flyback employs two control loops that control the gate of the MOSFET, represented in Figure 7a by the green block, and shown in detail in Figure 8. The inner control is a digitally implemented current mode controller implementing quasi-resonant switching. In this case, the internal current control (curved red line) regulates the amount of energy that is allowed to flow out of the charger and, in that way, sets a limit on the maximum continuous output current, typically 1 A or 2 A. The outer voltage feedback (curved blue line) controls the maximum output voltage of the charger and, accordingly, turns the charger on/off. The maximum output voltage is limited to the nominal voltage expected by e-bike's battery management system, which is typically 24, 36, and 48 volts. A Type-2 control is implemented and is shown on the left side of the control IC in Figure 8, including the current mode controller and gate driver [49].

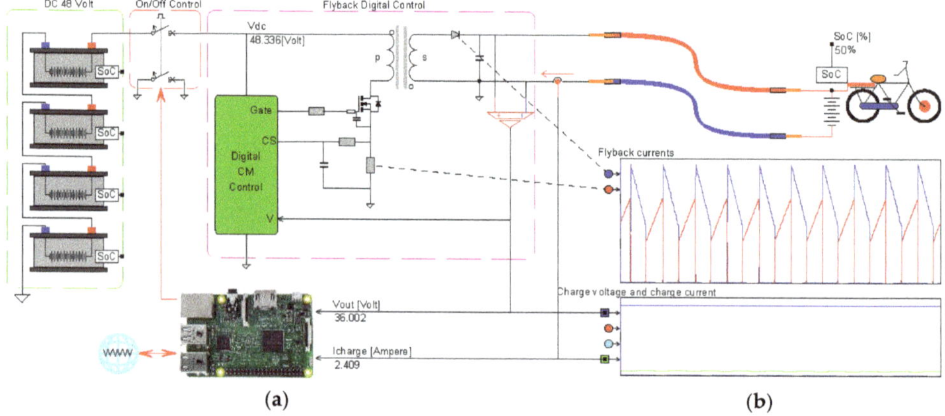

Figure 7. (**a**) Flyback converter for e-bike battery charging with a digital current mode control and charging current monitoring and logging. (**b**) Simulation results show the primary (red) and secondary (blue) current as well as the charge voltage (V_{out} = 36 Volt) and charge current (I_{ch} = 2.4 Ampere) for the 48-volt DC grid, and operating in a continuous conduction mode.

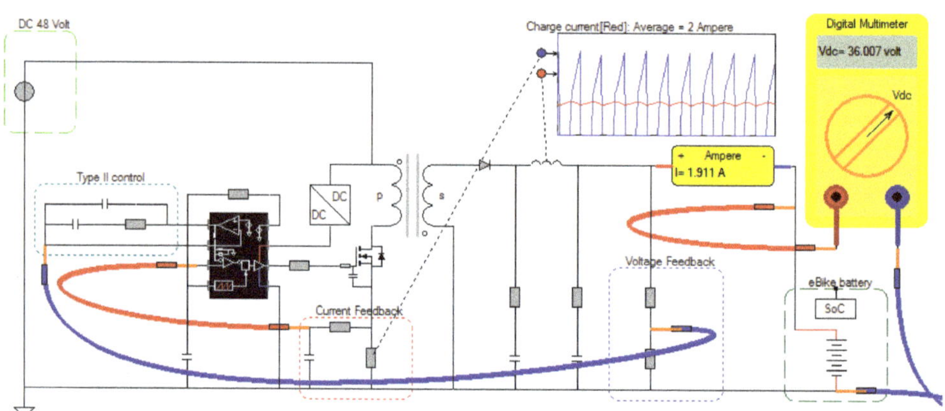

Figure 8. Current mode control of the flyback converter using an outer voltage loop and an inner current loop.

5.2. Design and Simulation of the Flyback DC Charger

The flyback converter is designed based on the design procedure for a quasi-resonant operation [10]. Figure 7b shows the simulation of the e-bike charger in Caspoc, where only a single leg flyback is shown for simplicity [50,51]. The multi-level modelling method is employed, which allows the modelling of both the power electronics circuit and the hybrid control consisting of the inner current and outer voltage control loop [52]. The simulation takes into account the delays of the digital feedback compensation as well as the delays caused by switching of the MOSFET. Voltage overshoot caused by the non-coupled parasitic winding inductance is taken into account in the simulation, but not shown here in Figure 7b for clarity.

Figure 7b shows the primary and secondary currents for a duty cycle of 50% in a continuous conduction mode for a switching frequency f_s = 100 kHz and turns the ratio of N_p:N_s = 48:36. The scope clearly shows the difference in the voltage level on the primary side of 48 V, and secondary side of 36 V, and the charging current of 2.4 A, which is measured on the second scope.

5.3. Safety and Monitoring

As long as no e-bike is connected, the mechanical on/off switch on the flyback's primary side (Figure 6a) is in the off position. As soon as an e-bike battery is connected to the flyback's secondary side, the voltage is monitored, and the switch is set to the 'on' position. This subsequently powers the flyback converter and its internal control. If either the connection with the e-bike's battery is removed or the charging current drops below the minimum charging current threshold, the outer control turns off the flyback converter by setting the On/Off control to its off position.

5.4. Hardware Realization and Losses

Figure 9a shows the hardware of the flyback DC charger [48]. On the primary side, IRFS4321PBF MOSFETs with low on-state resistance $R_{ds(on)} = 15$ mΩ and, on the secondary side, MBR10150 schottky diodes with negligible reverse recovery losses are used. The primary current is sensed using a low ohmic sense resistor placed between the MOSFET source and ground, which is filtered using a low pass first-order filter. In this way, the maximum power is limited by the maximum primary current. The output voltage is measured and, after filtering, sampling is input to the current mode controller. The feedback amplifier with a Type-2 compensation network is replaced by its digital equivalent 2-Pole, 2-Zero compensator [2p2z].

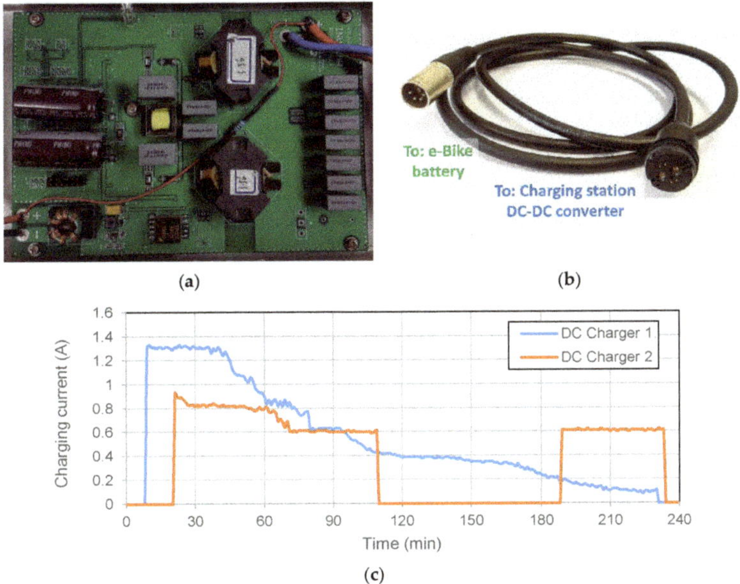

Figure 9. (a) Interleaved flyback converter for DC e-bike charging. (b) The cable used for DC charging. (c) Measured charging current of two e-bikes in which each is charged using the flyback based DC e-bike charger.

For the converter losses, the worst-case operating point is at the lowest output voltage of 24 V and maximum power of 100 W, which corresponds to a current of 4.2 A. For the dual interleaved flyback converter, each converter delivers 50 W and 2.1 A current. The MPR10150 diode has a maximum forward voltage $V_f = 0.6$V at 2.1A, conducts only 50% of the time, and, therefore, has $P_d = 50\%$ (2.1A)(0.6V) = 0.62 W losses. The average input current for each converter and an input voltage of 48V is roughly $I_{in} = 1A$. Considering a nominal 50% duty cycle, this will give a peak input current $I_{pk} = 4$ A and an RMS value of $I_{in(rms)} = 1.7$ A. This results in a maximum conduction loss of $(I_{in(rms)})^2 R_{ds(on)} = 44$ mW. As a rule of thumb, the switching losses are of the same magnitude of maximum and twice the conduction losses, which gives a total of conduction and switching losses of $P_{sw} = 3(44$ mW$) = 132$ mW. Cooling via the PCB is, hence, sufficient in this case.

Figure 9b shows the custom-designed cable that is used to connect the e-bike to the output of the flyback converter in the charging station. The right-side plug is magnetic and, hence, easily attaches to the station, while the left side plug attaches to the e-bike battery. If required, a custom-designed adapter is made for the e-bike side plug based on the manufacturer-specific connector on the user's e-bike. This points to the need for standardization in DC e-bike charging and plugs similar to what was achieved with CCS and CHAdeMO in electric car charging [53].

Figure 9c shows the charging current of two e-bikes in which each was charged using the flyback based DC e-bike charger. For charger 1, the battery gets charged according to the Constant Current, Constant Voltage (CC-CV) principle beginning at a current of 1.3 A, and then slowly reducing to zero. For charger 2, the battery charging begins in the CV region due to the relatively high state of charge (SOC) of the e-bike battery. It can be observed how the e-bike is disconnected at 120 min and then connected again after about an hour.

5.5. Battery Connection Communication

The internal digital control employs an automated voltage level detection to allow the direct connection to either a 24 V, 36 V, or 48 V battery. An automated output voltage level detection of the battery voltage is achieved through the outer control loop that compares the output voltage to the reference value. The outer control loop subsequently regulates the amplitude of the primary current pulse through the inner loop. The Raspberry Pi is used to control the flyback operation and to monitor the output voltage and charging current. Connected to the Internet, it exchanges and logs information with a server operating in the cloud.

6. Wireless E-Bike Charging

Wireless charging provides the most convenient and safe experience for the e-bike user. The cyclists do not need to bring along cables and power adapter because the charging process is facilitated through the bike kickstand, as shown in Figure 10. The developed charging system is started once the bike is parked on the appointed parking spot, which is a 30 × 30 cm tile underneath the solar charging station. On the tile, the users have the possibility to park their bike in any position, which makes the wireless charging convenient. On top of that, the wireless charging has intrinsic galvanic insolation, which does not require the users to touch any conductive parts such as cables and connectors that might become dangerous, especially in wet weather conditions.

6.1. Wireless Power Transfer via Resonant Circuit

Figures 10, 11a and Table 3 shows the schematic, circuit diagram, and specifications of the wireless power transfer system for the e-bike that works based on inductive power transfer through magnetic resonance. In the developed system, the transmitter coil is located under the charging tile, and it is formed by a U-shaped ferromagnetic core with a winding located at the center. On the other hand, the receiver coil consists of the double kickstand of the e-bike. This coil has a similar magnetic circuit as the transmitter coil but, in this case, the ferromagnetic core is closer to a V-shape to resemble the structure of commercial double kickstands.

Once the bike is parked over the tile, the two coils become coupled, which is equivalent to saying that the magnetic circuit becomes closed, and the charging process is ready to start. In a traditional transformer, the coils are strongly coupled because they are wounded around the same ferromagnetic core, and the airgap's order of magnitude is not comparable to the core's dimensions. On the other hand, when the two coils become coupled, the proposed magnetic circuit has two airgaps of about

5 mm, which are less than one order of magnitude smaller than the cross-sectional area of the core. Therefore, in this case, the equivalent transformer has a high leakage inductance and, consequently, is loosely coupled.

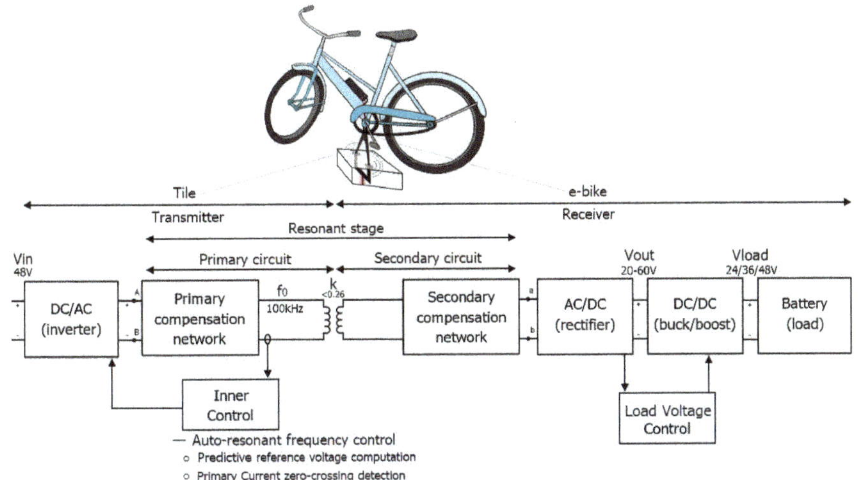

Figure 10. Block diagram of the wireless power transfer system for the e-bike.

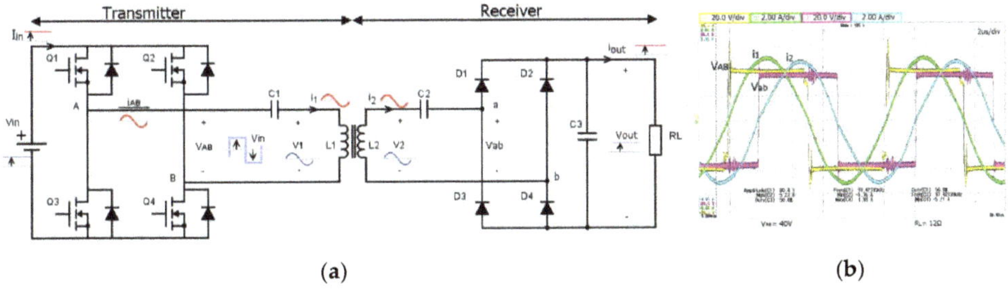

(a) (b)

Figure 11. (a) Simplified circuit of the e-bike wireless charging system. (b) Waveforms of the primary and secondary voltages and currents.

Table 3. Values of the components of the e-bike wireless charging system.

V_{in} (V)	V_{out} (V)	V_{load} (V)	P_{out} (W)	f_0 (kHz)	η_{max} (%)	**MOSFETs & Diodes**
48	20–60	24/36/48	200	90–110	89.2	IPP030N10N5 BYW29-200
k	L_1 (μH)	L_2 (μH)	C_1 (nF)	C_1 (nF)	R_1 (Ω)	R_1 (Ω)
0.28	67.7	46.3	35.9	52.3	0.11	0.16

The capacitors C_1 and C_2 compensate the coils' reactive power and form a resonant circuit together with the inductor coils L_1 and L_2, which is tuned to the resonant frequency f_0. The estimation of C_1 and C_2 and the Kirchhoff voltage law equations of the circuit schematic can be found in Equations (9) and (10).

$$\omega_0 = 2\pi f_0 C_1 = \frac{1}{\omega_0^2 L_1} \quad C_2 = \frac{1}{\omega_0^2 L_2} \tag{9}$$

$$\frac{4}{\pi} V_{in} = \left(j\omega L_1 + \frac{1}{j\omega C_1}\right)I_1 + j\omega M I_2$$
$$0 = \left(\frac{8}{\pi^2} R_L + j\omega L_2 + \frac{1}{j\omega C_2}\right)I_2 + j\omega M I_1 \tag{10}$$

As shown in Figure 10, the wireless charging system draws power directly from the 48 V DC nano-grid. The DC input voltage is then inverted such that a high-frequency voltage supplies the transmitter resonant circuit, which is also called the primary circuit, and the current produces a magnetic field that links to the secondary coil. The high-frequency voltage induced in the secondary circuit is then rectified, and its value is regulated via a DC/DC converter to be supplied correctly to the battery.

6.2. Variable Frequency for Misalignment Tolerance

Since the cyclist has flexibility in parking the bike, the coupling between the two coils can vary. Moreover, depending on the state of charge of the battery, the loading condition also changes during the charging process. On top of that, the value of the circuit components can change because of an increase in temperature or degradation with time. All these factors make the resonant frequency of the system vary from the designed value. To overcome this problem, the inverter's operating frequency is set by the inner control loop that tracks the natural resonant frequency of the system. For this reason, it is called auto-resonant frequency control. It sets the soft switching of the inverter by predicting the primary current zero-crossing depending on the current slope. This control circuit is simple, analog, fast, and it automatically adapts the operating frequency to the optimum value in a few periods.

6.3. Communication with the Bike and Foreign Object Identification

The communication between the e-bike and the charging station allows the start-up, shut-down, and foreign object detection. It is realized through backscatter modulation in the power line through amplitude shift keying (ASK) modulation. In this way, the information can be reflected by one side to the other by modulating the voltage of the resonant circuit between a low and high value with a frequency lower than the operating one (about 1 kHz). To send information via ASK from the e-bike to the station, a resistor in series with a switch connected across the output DC voltage is used. When the switch is in an open or close position, ASK assumes a high or low value (due to a voltage drop across the resistor), respectively. These binary values can be organized in sets of bits that form the messages of the communication system. Once a message is received, it is demodulated and interpreted for execution.

The wireless charging system has to be able to detect and stop the charging process if a foreign object is placed on the surface of the transmitter coil. This is because the magnetic field can potentially heat the object through eddy currents, and could lead to fire or injury. This is because either the charging process is not started or the ongoing charging process is stopped if a foreign object is detected. The first scenario is avoided because the charging begins only if proper communication messages are sent by the e-bike. On the other hand, efficiency measurements of the power transfer can indicate if a foreign object is receiving part of the transferred power, which results in a lower efficiency level than expected.

6.4. Experimental Realization

Figure 12a shows the picture of the wireless e-bike charger and the laboratory setup used as proof-of-concept to test its operation. The realization of the wireless charging for the e-bike at the solar station is shown in Figure 12b. Table 3 shows the parameter values of the various circuit components of the resonant circuit shown in Figure 11a. For the inverter, the 100V IPP030N10N5 MOSFETs switches are used, and, for the rectifier, 200V BYW29-200 diodes are used.

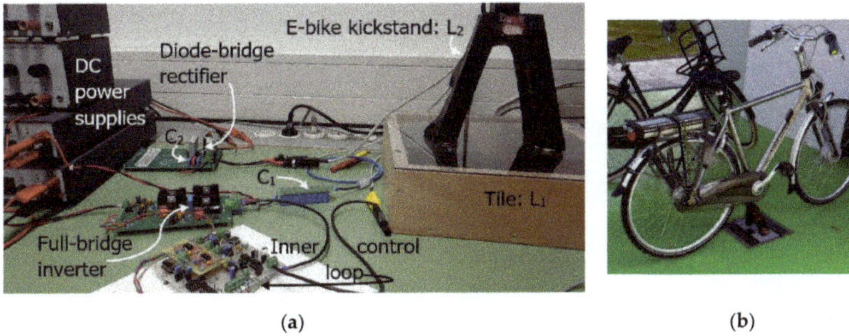

Figure 12. (**a**) Laboratory setup of the e-bike wireless charging system and (**b**) e-bike with primary coil placed below the tile in the station and secondary coil integrated into the kickstand.

Figure 11b shows the voltage and current waveforms of the primary and secondary circuits. In this case, the zero-current switching (ZCS) of unity power factor is achieved because the voltage and current V_{AB} and I_1 are in phase. Therefore, the power transfer is maximized, and the ZCS of the inverter is achieved. The auto-resonant frequency control can also achieve zero-voltage switching (ZVS) depending on the gain given to the predictive zero-crossing current detection. With ZVS, the power factor is slightly less than the unity because the current I_1 is lagging the voltage V_{AB}. In Reference [54], the advantages of ZCS and ZVS are analyzed in detail by considering this e-bike charging system. The measured efficiency from the 48 V DC nano-grid to the EV bike batteries is 89.2% at the maximum coupling condition and with an output power of 230 W.

6.5. Ongoing Development

Currently, the functionality of the e-bike wireless charger has been proved in the laboratory. After this, measurements of the electromagnetic field radiated by the charger are going to be performed to verify the compliance to the International Commission on Non-Ionizing Radiation Protection (ICNIPR) that ensures safety to human beings [55]. The radiated magnetic field of the charger is expected to be within the limits for general public (27 µT) since the coils are placed on top of each other, and there is no gap in the free air. This is unlike in electric cars, where the coils have a flat arrangement with a larger gap [56]. Moreover, the rated power of this charger is considerably lower than the 3.3 kW minimum for electric cars as per SAE J2954 [57]. Hence, a lower radiated magnetic flux can be expected. Second, a kickstand with structurally integrated magnetics is being developed, which is sturdier for a longer lifetime.

7. Environment Integrated PV System

The solar panels, battery storage, and the AC, DC, and wireless charging are combined together to form an Environment Integrated PV system (EIPV) built on the university campus, as shown in Figure 2. Three cabinets of 1.39 m by 0.72 m are located inside the EIPV, which are used for storing the battery and the associated electronics for the inverter and MPPT converter and the DC e-bike charger and control circuitry, respectively. Integrating all the electronics and batteries inside the EIPV saves approximately 3 m^3 space that would have otherwise be required for external cabinets to house all the electronics and batteries. The key advantage of the EIPV is, therefore, the mechanical and electrical integration of all components resulting in a single structure that combines aesthetics, modularity, safety, functionality, ergonomics, and usability.

Figure 13a shows the solar MPPT converter, bidirectional inverter, grid islanding device, control, and protection circuity for both devices. Figure 13b shows the DC-DC converters for the e-bike DC charging, charging measurement circuit, and the Raspberry Pi central controller responsible for communicating with all the devices. The Raspberry Pi also reads data from all devices like the VICTRON system, DC chargers, and weather station and logs them centrally into an Internet server [58]. The charging current and voltage are displayed on the monitor inside the e-bike station for user convenience.

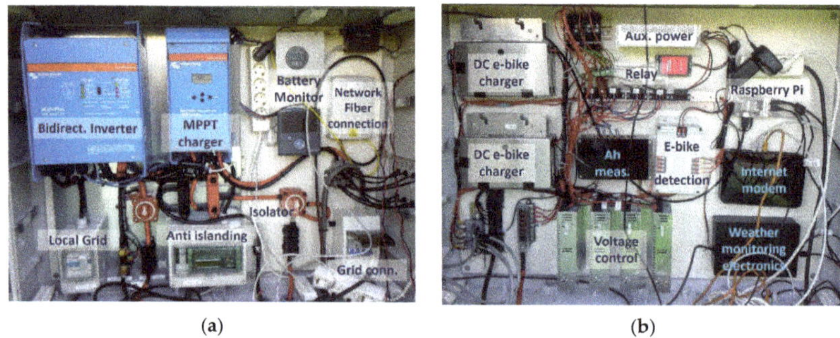

(a)   (b)

Figure 13. (**a**) Grid Inverter, PV MPPT charger, protection, and monitoring circuit. (**b**) DC-DC converter, control, and protection for DC charging of e-bike.

7.1. Energy Yield of the PV System and Load

Figure 14 shows the measured monthly energy yield of a 2.6 kW PV system for one year over the period of October 2018 to September 2019. A total of 2378 kWh of PV energy is produced in this period, corresponding to a daily average of 6.5 kWh/day. The extreme differences in the PV generation between the seasons can be seen with 40 kWh generation in December compared to 315 kWh in June, up to an eight times difference. In terms of daily energy yield, there is as much as a 25 times difference, varying from 0.64 kWh/day to 15.4 kWh/day. It is also important to note the difference in yield between Figure 14, and Figure 5b due primarily to the difference in the meteorological conditions of 2013 and 2018/2019.

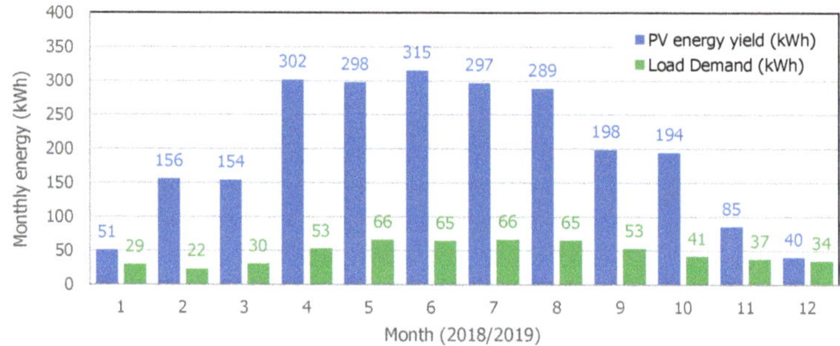

Figure 14. Monthly energy yield (in kWh) of the 2.6 kW PV system for the period of October 2018 to September 2019 and the corresponding load demand including losses (in kWh).

In terms of load demand, two e-bikes from the electrical engineering faculty are regularly charged at the location with occasional demand from other e-bikes, e-scooters, and the Twizy. The annual demand of the station was much lower than the theoretical analysis and was found to be 561 kWh/year or 1.5 kWh/day on average. The lower demand was due to a much lower baseload (limited use of light, converters going into sleep mode) and a lesser number of e-bikes than anticipated. The load also exhibited a seasonal variation primarily due to variation in usage of e-bikes with higher e-bike usage and charging in the summer.

From an economic perspective, an annual yield of 2378 kWh of PV energy results in a revenue of ~595€/year, assuming a net-metered feed-in tariff of 0.25 €/kWh. Charging a 500-Wh e-bike battery costs about 0.125 €/charge. Due to the custom design of the station, the net cost of all the electronics, including the PV, battery, and chargers was approximately 15,000 €, which resulted in a payback period of ~25 years. It must be noted that this does not include civil material costs of constructing the station and the cost of hours for research and development, which are significant [9]. To reduce the costs, the options can be to provide only AC charging, not including a battery, installing the PV on the rooftop, and using a single converter for both MPPT and grid feeding.

7.2. Power Management of the Battery

The integrated battery storage can be controlled with numerous power management strategies based on the PV generation and grid conditions. One such strategy that has been implemented currently is shown below.

- The PV power is primarily used to provide the e-bike charging load and the baseload
- If the available PV power is less than the load demand, then the demand is first supplied by the battery. If the battery becomes empty, then the load is supplied by the grid.
- If the available PV power is more than the demand, then the excess PV power is first used to charge the battery. If the battery is nearly full, then the PV power is fed to the AC grid.
- When the battery SOC > 50%, the battery and PV power together feed at least 400 W to the AC grid in order to utilize the battery in order to store the solar power the next day.

Based on this power management, Figure 15 shows the measured power profiles of the PV, battery, grid, and the battery SOC over one week in May with a few cloudy days. First, it can be observed that the solar power is used to supply the load on sunny days, feed at least 400 W to the AC grid, and to charge the battery from 50% to 100% SOC. Second, there is a dip in the PV generation in the afternoon due to shading from the nearby faculty building, as seen in Figure 5. Third, when the solar production is low/zero (especially in the evening and night), the battery discharges and feeds power to the AC grid. Lastly, when there is not enough solar production and if the battery SOC ≤ 50%, no power is supplied to the AC grid.

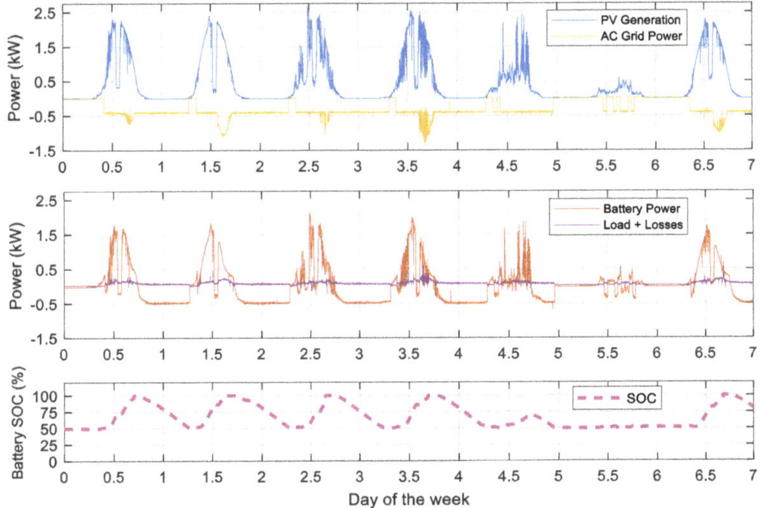

Figure 15. Measurement over one week in May: (**Top**) the generated PV power and power fed to the grid. (**Middle**) battery charging power (positive), discharging power (negative), the load, and power losses. (**Bottom**) SOC of battery.

8. Conclusions

A 2.6 kW$_p$ solar powered charging station for e-bikes and e-scooters has been designed and installed that offers AC, DC, and wireless charging. It has an integrated storage with a usable capacity of 9.5 kWh that can provide both grid-connected and off-grid operation using a hybrid bidirectional inverter. The station has a 48 V DC nano-grid that is used for power exchange between the PV, EV, battery, and the AC grid. The 100 W DC charging and 200 W wireless charging systems are unique in that the user can charge the e-bike without requiring an AC charging adapter. The AC charging system provides up to 3.7 kW charging power, which is sufficient for a small electric car like the Renault Twizy.

The PV orientation was optimized to a tilt angle at 51° and facing south to increase PV yield in the winter month of December while not compromising on the annual yield significantly (<5%). 3D modelling using Sketchup is used to determine the shading due to nearby buildings to make an accurate estimation of the yield. In the observed period of 2018/2019, 2378 kWh of PV energy is produced, which corresponded to a daily average of 6.5 kWh/day. At the same time, the seasonal variation in irradiance caused up to 25 times variation in daily yield from 0.64 kWh/day to 15.4 kWh/day.

The DC charging system uses current-mode controlled flyback converters to charge 24–48 V e-bike batteries from the 48 V DC nano-grid. A custom-designed DC cable can be used to connect the e-bike batteries of different manufacturers to the station. On the other hand, the wireless charging system uses two windings on a U-shaped and V-shaped ferromagnetic core, with one placed under the tile of the charging station and the other integrated into the bike kickstand, respectively. A unique auto-resonant frequency control and amplitude shift keying modulation are implemented for misalignment tolerance and foreign object detection, which are crucial for practical usage.

The environmentally-integrated PV system cohesively integrates the mechanical, structural, and electrical components in a single unit, which saves space and provides aesthetics, modularity, safety, ergonomics, and convenience. The charging station, including its weather station, can be remotely controlled and monitored using a Raspberry Pi.

Author Contributions: Conceptualization, G.R.C.M., P.V.D., P.B., and O.I. Methodology, validation, formal analysis & investigation, G.R.C.M., P.V.D., F.G., and A.J. Writing—original draft preparation, G.R.C.M. Writing—review and editing, G.R.C.M., P.V.D., and F.G. Visualization, G.R.C.M., P.V.D., and F.G. Supervision, G.R.C.M., P.V.D., P.B., and O.I. Project administration, P.V.D. Funding acquisition, P.V.D., P.B., and O.I. All authors have read and agreed to the published version of the manuscript.

Funding: The Delft Infrastructure & Mobility Initiative, 3E fonds, Climate-KIC funded this research. The authors thank them for their financial support.

Acknowledgments: The authors would like to acknowledge (in alphabetical order) Bart Roodenburg, Gireesh Nair, Harrie Olsthoorn, Joris Koeners, Miro Zeman, Sacha Silvester, Tim Velzeboer, and Yunpeng Zhao of the Delft University of Technology, the Netherlands for their technical guidance and support, and Junyin Gu from Involar for the in-kind support of hardware components.

Conflicts of Interest: The authors declare no conflict of interest.

References

1. Messagie, M.; Boureima, F.-S.S.; Coosemans, T.; Macharis, C.; Mierlo, J. Van A range-based vehicle life cycle assessment incorporating variability in the environmental assessment of different vehicle technologies and fuels. *Energies* **2014**, *7*, 1467–1482. [CrossRef]
2. Moro, A.; Lonza, L. Electricity carbon intensity in European Member States: Impacts on GHG emissions of electric vehicles. *Transp. Res. Part D Transp. Environ.* **2018**. [CrossRef] [PubMed]
3. *Dataset nrg_105a: Supply, Transformation and Consumption of Electricity—Annual Data, Statistical Office of the European Union (EUROSTAT)*; Eurostat: Luxembourg, 2016.
4. Hamilton, C.; Gamboa, G.; Elmes, J.; Kerley, R.; Arias, A.; Pepper, M.; Shen, J.; Batarseh, I. System architecture of a modular direct-DC PV charging station for plug-in electric vehicles. In Proceedings of the Annual Conference on IEEE Industrial Electronics Society, Glendale, AZ, USA, 7–10 November 2010; pp. 2516–2520.

5. Carli, G.; Williamson, S.S. Technical Considerations on Power Conversion for Electric and Plug-in Hybrid Electric Vehicle Battery Charging in Photovoltaic Installations. *IEEE Trans. Power Electron.* **2013**, *28*, 5784–5792. [CrossRef]

6. Chandra Mouli, G.R.; Bauer, P.; Zeman, M. System design for a solar powered electric vehicle charging station for workplaces. *Appl. Energy* **2016**, *168*, 434–443. [CrossRef]

7. Birnie, D.P. Solar-to-vehicle (S2V) systems for powering commuters of the future. *J. Power Sour.* **2009**, *186*, 539–542. [CrossRef]

8. Lai, C.S.; McCulloch, M.D. Levelized cost of electricity for solar photovoltaic and electrical energy storage. *Appl. Energy* **2017**. [CrossRef]

9. Chandra Mouli, G.R.; Leendertse, M.; Prasanth, V.; Bauer, P.; Silvester, S.; van de Geer, S.; Zeman, M. Economic and CO2 Emission Benefits of a Solar Powered Electric Vehicle Charging Station for Workplaces in the Netherlands. In Proceedings of the IEEE Transportation Electrification Conference and Expo (ITEC), Dearborn, MI, USA, 27–29 June 2016; pp. 1–7.

10. Chandra Mouli, G.R.; Schijffelen, J.; Van Den Heuvel, M.; Kardolus, M.; Bauer, P. A 10 kW Solar-Powered Bidirectional EV Charger Compatible with Chademo and COMBO. *IEEE Trans. Power Electron.* **2019**, *34*, 1082–1098. [CrossRef]

11. Fairley, P. China's cyclists take charge: Electric bicycles are selling by the millions despite efforts to ban them. *IEEE Spectr.* **2005**, *42*, 54–59. [CrossRef]

12. *Mobiliteit in Cijfers Tweewielers 2017–2018, Stichting BOVAG-RAI Mobiliteit, Amsterdam*; Amsterdam Foundation: Amsterdam, The Netherlands, 2017.

13. Apostolou, G.; Reinders, A.; Geurs, K. An Overview of Existing Experiences with Solar-Powered e-Bikes. *Energies* **2018**, *11*, 2129. [CrossRef]

14. Chandra Mouli, G.R.; van Duijsen, P.; Velzeboer, T.; Nair, G.; Zhao, Y.; Jamodkar, A.; Isabella, O.; Silvester, S.; Bauer, P.; Zeman, M. Solar Powered E-Bike Charging Station with AC, DC and Contactless Charging. In Proceedings of the European Conference on Power Electronics and Applications (EPE'18 ECCE Europe), Riga, Latvia, 17–21 September 2018.

15. Mesentean, S.; Feucht, W.; Mittnacht, A.; Frank, H. Scheduling Methods for Smart Charging of Electric Bikes from a Grid-Connected Photovoltaic-System. In Proceedings of the 2011 UKSim 5th European Symposium on Computer Modeling and Simulation, Madrid, Spain, 16–18 November 2011; pp. 299–304.

16. Fuentes, M.; Fraile-Ardanuy, J.; Risco-Martín, J.L.; Moya, J.M. Feasibility Study of a Building-Integrated PV Manager to Power a Last-Mile Electric Vehicle Sharing System. *Int. J. Photoenergy* **2017**. [CrossRef]

17. Zhang, Z.; Gercek, C.; Renner, H.; Reinders, A.; Fickert, L. Resonance instability of photovoltaic E-bike charging stations: Control parameters analysis, modeling and experiment. *Appl. Sci.* **2019**, *9*, 252. [CrossRef]

18. Thomas, D.; Klonari, V.; Vallee, F.; Ioakimidis, C.S. Implementation of an e-bike sharing system: The effect on low voltage network using pv and smart charging stations. In Proceedings of the 2015 International Conference on Renewable Energy Research and Applications, ICRERA 2015, Palermo, Italy, 22–25 November 2015; pp. 572–577.

19. Kulshrestha, P.; Wang, L.; Chow, M.Y.M.-Y.; Lukic, S. Intelligent energy management system simulator for PHEVs at municipal parking deck in a smart grid environment. In Proceedings of the 2009 IEEE Power and Energy Society General Meeting, PES '09, Calgary, AB, Canada, 26–30 July 2009; pp. 1–6.

20. Zhou, W.; Wattenberg, M.U.S. Design Considerations of a Single Stage LLC Battery Charger. In Proceedings of the PCIM Europe; International Exhibition and Conference for Power Electronics, Intelligent Motion, Renewable Energy and Energy Management, Nuremberg, Germany, 7–9 May 2019.

21. Nguyen, C.-L.; Primiani, P.; Viglione, L.; Woodward, L. A Low-Cost Battery Charger Usable with Sinusoidal Ripple-Current and Pulse Charging Algorithms for E-Bike Applications. In Proceedings of the 2019 IEEE 28th International Symposium on Industrial Electronics (ISIE), Vancouver, BC, Canada, 12–14 June 2019; pp. 2085–2090.

22. Chandra Mouli, G.R.; Bauer, P.; Zeman, M. Comparison of system architecture and converter topology for a solar powered electric vehicle charging station. In Proceedings of the International Conference on Power Electronics and ECCE Asia (ICPE-ECCE Asia), Seoul, Korea, 1–5 June 2015; pp. 1908–1915.

23. Vega-Garita, V.; Ramirez-Elizondo, L.; Chandra Mouli, G.R.; Bauer, P. Review of residential PV-storage architectures. In Proceedings of the 2016 IEEE International Energy Conference (ENERGYCON), Leuven, Belgium, 4–8 April 2016; pp. 1–6. [CrossRef]

24. Pellitteri, F.; Boscaino, V.; Di Tommaso, A.O.; Genduso, F.; Miceli, R. E-bike battery charging: Methods and circuits. In Proceedings of the 2013 International Conference on Clean Electrical Power (ICCEP), Alghero, Italy, 11–13 June 2013; pp. 107–114.

25. Pellitteri, F.; Di Tommaso, A.O.; Miceli, R. Investigation of inductive coupling solutions for E-bike wireless charging. In Proceedings of the Universities Power Engineering Conference, Computer Society, Stoke on Trent, UK, 1–4 September 2015.

26. Genco, F.; Longo, M.; Patrizia Livrieri, A.T. Wireless Power Transfer System Stability Analysis for E-bikes Application. In Proceedings of the AEIT International Conference of Electrical and Electronic Technologies for Automotive (AEIT AUTOMOTIVE), Torino, Italy, 2–4 July 2019.

27. Joseph, P.K.; Elangovan, D.; Arunkumar, G. Linear control of wireless charging for electric bicycles. *Appl. Energy* **2019**, *255*, 113898. [CrossRef]

28. Livreri, P.; Di Dio, V.; Miceli, R.; Pellitteri, F.; Galluzzo, G.R.; Viola, F. Wireless battery charging for electric bicycles. In Proceedings of the 2017 6th International Conference on Clean Electrical Power (ICCEP), Santa Margherita Ligure, Italy, 27–29 June 2017; pp. 602–607.

29. Pellitteri, F.; Ala, G.; Caruso, M.; Ganci, S.; Miceli, R. Physiological compatibility of wireless chargers for electric bicycles. In Proceedings of the 2015 International Conference on Renewable Energy Research and Applications, ICRERA 2015, Palermo, Italy, 22–25 November 2015; pp. 1354–1359.

30. Iannuzzi, D.; Rubino, L.; Di Noia, L.P.; Rubino, G.; Marino, P. Resonant inductive power transfer for an E-bike charging station. *Electr. Power Syst. Res.* **2016**. [CrossRef]

31. Chen, Y.; Kou, Z.; Zhang, Y.; He, Z.; Mai, R.; Cao, G. Hybrid Topology With Configurable Charge Current and Charge Voltage Output-Based WPT Charger for Massive Electric Bicycles. *IEEE J. Emerg. Sel. Top. Power Electron.* **2018**, *6*, 1581–1594. [CrossRef]

32. Kindl, V.; Pechanek, R.; Zavrel, M.; Kavalir, T. Inductive coupling system for E-bike wireless charging. In Proceedings of the 12th International Conference ELEKTRO 2018, Mikulov, Czech Republic, 21–23 May 2018; pp. 1–4.

33. Coppola, M.; Cennamo, P.; Dannier, A.; Iannuzzi, D.; Meo, S. Wireless Power Transfer circuit for e-bike battery charging system. In Proceedings of the 2018 IEEE International Conference on Electrical Systems for Aircraft, Railway, Ship Propulsion and Road Vehicles and International Transportation Electrification Conference, ESARS-ITEC, Nottingham, UK, 7–9 November 2018.

34. Lee, J.; Lee, S.; Kwak, M.; Yeom, G. Seung-Hwan Lee Development of a Wireless Power Supply System for an E-Bike. In Proceedings of the 10th International Conference on Power Electronics and ECCE Asia (ICPE 2019—ECCE Asia), Busan, Korea, 27–30 May 2019.

35. Joseph, P.K.; Elangovan, D. A review on renewable energy powered wireless power transmission techniques for light electric vehicle charging applications. *J. Energy Storage* **2018**, *16*, 145–155. [CrossRef]

36. Nunes, P.; Figueiredo, R.; Brito, M.C. The use of parking lots to solar-charge electric vehicles. *Renew. Sustain. Energy Rev.* **2016**, *66*, 679–693. [CrossRef]

37. Mackay, L.; Hailu, T.G.; Chandra Mouli, G.R.; Ramirez-Elizondo, L.; Ferreira, J.A.; Bauer, P. From DC Nano- and Microgrids Towards the Universal DC Distribution System—A Plea to Think Further Into the Future. In Proceedings of the 2015 IEEE Power & Energy Society General Meeting, Denver, CO, USA, 26–30 July 2015; pp. 1–5.

38. Victron Energy. *Datasheet—MultiPlus 48/3000/35-16; BlueSolar Charger MPPT 150/85*; Victron Energy: Almere, The Netherlands, 2019.

39. Lufft. Datsheet. WS503-UMB Smart Weather Sensor. Available online: https://www.lufft.com/products/compact-weather-sensors-293/ws503-umb-smart-weather-sensor-1837/productAction/outputAsPdf/ (accessed on 1 July 2020).

40. Chao, D.C.-H.; van Duijsen, P.J.; Hwang, J.J.; Liao, C.-W. Modeling of a Taiwan fuel cell powered scooter. In Proceedings of the 2009 International Conference on Power Electronics and Drive Systems (PEDS), Taipei, Taiwan, 2–5 November 2009; pp. 913–919.

41. CESAR Database, Koninklijk Nederlands Meteorologisch Instituut (KNMI). Available online: http://www.knmi.nl (accessed on 1 July 2020).

42. Paulescu, M.; Paulescu, E.; Gravila, P.; Badescu, V. Weather Modeling and Forecasting of PV Systems Operation. *Green Energy Technol.* **2013**, *103*. [CrossRef]

43. Braun, J.E.; Mitchell, J.C. Solar geometry for fixed and tracking surfaces. *Sol. Energy* **1983**, *31*, 439–444. [CrossRef]

44. Reda, I.; Andreas, A. Solar position algorithm for solar radiation applications. *Sol. Energy* **2004**, *76*, 577–589. [CrossRef]

45. Jakhrani, A.Q.; Othman, A.K.; Rigitand, A.R.H.; Samo, S.R. Comparison of solar photovoltaic module temperature models. *World Appl. Sci. J.* **2011**, *14*, 1–8.

46. Isabella, O.; Nair, G.G.; Tozzi, A.; Castro Barreto, J.H.; Chandra Mouli, G.R.; Lantsheer, F.; van Berkel, S.; Zeman, M. Comprehensive modelling and sizing of PV systems from location to load. *MRS Proc.* **2015**, *1771*, 1–7. [CrossRef]

47. Smets, A.; Jäger, K.; Isabella, O.; van Swaaij, R.; Zeman, M. *Solar Energy: The Physics and Engineering of Photovoltaic Conversion, Technologies and Systems*; UIT: Cambridge, UK, 2016; ISBN 9780203841464.

48. *Involar Micro-Inverters*; Involar Corporation Ltd.: Pudong District, Shanghai, China.

49. Ridley, R.B. *Power Supply Design, Volume 1: Control*; Ridley Designs: Camarillo, CA, USA, 2012; ISBN 9780983318002.

50. *Caspoc*; Simulation Research: Alphen aan den Rijn, The Netherlands, 2019; Available online: www.caspoc.com (accessed on 1 July 2020).

51. Bauer, P.; Van Duijsen, P.J. Challenges and Advances in Simulation. In Proceedings of the IEEE 36th Conference on Power Electronics Specialists, Recife, Brazil, 12–16 June 2005; pp. 1030–1036.

52. Van Duijsen, P.; Woudstra, J.; Van Willigenburg, P. Educational setup for Power Electronics and IoT. In Proceedings of the 2018 19th International Conference on Research and Education in Mechatronics, REM 2018, Delft, The Netherlands, 7–8 June 2018.

53. Chandra Mouli, G.R.; Kaptein, J.; Bauer, P.; Zeman, M. Implementation of dynamic charging and V2G using Chademo and CCS/Combo DC charging standard. In Proceedings of the IEEE Transportation Electrification Conference and Expo (ITEC), Dearborn, MI, USA, 27–29 June 2016; pp. 1–6.

54. Grazian, F.; Duijsen, P.; Thiago, B.; Soeiro, P.B. Advantages and Tuning of Zero Voltage Switching in a Wireless Power Transfer System. In Proceedings of the IEEE PELS Workshop on Emerging Technologies: Wireless Power (WoW), London, UK, 17–23 June 2019.

55. Xu, H.; Wang, C.; Xia, D.; Liu, Y. Design of Magnetic Coupler for Wireless Power Transfer. *Energies* **2019**, *12*, 3000. [CrossRef]

56. Zeng, H.; Liu, Z.; Hou, Y.; Hei, T.; Zhou, B. Optimization of Magnetic Core Structure for Wireless Charging Coupler. *IEEE Trans. Magn.* **2017**. [CrossRef]

57. Society of Automotive Engineers (SAE). J2954—Wireless Power Transfer for LightDuty Plug-In/ Electric Vehicles and Alignment Methodology. *SAE Int.* **2019**. Available online: https://www.sae.org/standards/content/j2954_201605/ (accessed on 1 July 2020).

58. TU Delft Solar e-Bike Charging Station. Available online: http://solarpoweredbikes.tudelft.nl (accessed on 1 July 2020).

Article

Hybrid Microgrid Energy Management and Control Based on Metaheuristic-Driven Vector-Decoupled Algorithm Considering Intermittent Renewable Sources and Electric Vehicles Charging Lot

Tawfiq M. Aljohani, Ahmed F. Ebrahim and Osama Mohammed *

Energy Systems Research Laboratory, Department of Electrical and Computer Engineering, Florida International University, Miami, FL 33174, USA; Taljo005@fiu.edu (T.M.A.); aebra003@fiu.edu (A.F.E.)
* Correspondence: mohammed@fiu.edu; Tel.: +1-305-348-3040

Received: 27 March 2020; Accepted: 25 June 2020; Published: 2 July 2020

Abstract: Energy management and control of hybrid microgrids is a challenging task due to the varying nature of operation between AC and DC components which leads to voltage and frequency issues. This work utilizes a metaheuristic-based vector-decoupled algorithm to balance the control and operation of hybrid microgrids in the presence of stochastic renewable energy sources and electric vehicles charging structure. The AC and DC parts of the microgrid are coupled via a bidirectional interlinking converter, with the AC side connected to a synchronous generator and portable AC loads, while the DC side is connected to a photovoltaic system and an electric vehicle charging system. To properly ensure safe and efficient exchange of power within allowable voltage and frequency levels, the vector-decoupled control parameters of the bidirectional converter are tuned via hybridization of particle swarm optimization and artificial physics optimization. The proposed control algorithm ensures the stability of both voltage and frequency levels during the severe condition of islanding operation and high pulsed demands conditions as well as the variability of renewable source production. The proposed methodology is verified in a state-of-the-art hardware-in-the-loop testbed. The results show robustness and effectiveness of the proposed algorithm in managing the real and reactive power exchange between the AC and DC parts of the microgrid within safe and acceptable voltage and frequency levels.

Keywords: Energy management and control; particle swarm optimization (PSO); hybrid AC/DC microgrid; electric vehicle charging and discharging control; artificial physics optimization (APO)

---

1. Introduction

Microgrids are one of the promising solutions for smarter and more efficient energy operations. The recent growth of small-scale energy generations as well as the rapid progress of power electronics applications has increased the attention toward microgrids control and management issues in recent years. Moreover, concerns about the reduction of power plants immense contribution of greenhouse gasses (GHG) have shed light on microgrids importance and the role they could play to reduce the release of toxic gases to the environment [1,2]. Another aspect of its importance is the recent shift toward more transportation electrification, which require more electric vehicles (EVs) charging structures on the distribution system. However, the rise of microgrids with its dependence on intermittence renewable energy sources (RES) as well as stochastic EVs activities have underlined voltage stability and frequency control problems that must be carefully addressed for more safe and resilient operation. Specifically, uncoordinated large-scale integration of renewables sources, as well as rapid adoption of EVs with highly stochastic charging and discharging activities, will lead to detrimental consequences such as voltage collapse, power quality problems, frequency, and stability oscillations, to name a few. Therefore, proper control and operation of microgrids is required to allow coordinated control mechanisms while taking into consideration the heterogeneous mix of parameters corresponding to different attached power sources [3]. Research on microgrids operation and control has been widely considered in the literature. The authors of reference [4] analyzed various architecture, management, and control in the microgrid paradigm, while the authors of reference [5] presented a survey on various research that considered the integration of distributed energy resources with microgrids in different countries. Reference [6] investigated a decentralized energy control scheme for autonomous poly-generation microgrid topology to achieve proper management in case of malfunctioning of downstream parts. The authors of reference [7] examined a game theory, multi-agent-based microgrid energy management system with the coordination of the decentralized agents are employed via Fuzzy Cognitive Map (FCM). Besides, the authors of reference [8] presented a valuable review study on various hierarchical control schemes of microgrids on the primary, secondary, and tertiary control layers that aim to reduce the overall operation cost while improving the controllability and the reliability of microgrids.

Photovoltaic solar (PV) is one of the most advanced and reliable forms of renewable energy sources. However, the utilization of PV systems has yet to overcome many operational issues to be considered a thoroughly reliable and dispatchable source of energy for microgrids. The most critical issue with the consideration of PV systems is its intermittency throughout the day. Such shortages in the PV system's supply of energy could be compromised with increasing the level of energy transfer to the microgrid through EVs discharging, which is one of the main aspects this paper is investigating. The work on PV systems is one of the widely considered research topics in the past decades. Optimization problems have been well-developed to investigate and verify the control of PV systems taking into consideration its stochastic nature such as in [9,10]. The authors of reference [9] investigated the effect of changing cell's temperature and solar irradiance on the design of various DC-DC converter topologies which are widely used in PV systems. The authors of reference [10] proposed a new topology scheme for a photovoltaic dc/dc converter which can drastically enhance the efficiency of a PV system by assessing the PV's module characteristics. The authors of reference [11] proposed an algorithm that offers dynamic distributed energy resources control that includes PV systems, small-scale wind turbines, controllable loads, and energy storage devices. Furthermore, studies on PV systems covered a wide range of applications, where reference [12] proposed an energy management strategy with PV-powered desalination station that is coupled with DC microgrids.

Another side of consideration in this work is the relative impact of EVs integration on the hybrid microgrid operation. Reference [13] presents a linearization methodology to model real-time EVs activities on residential feeders based on the concept of Kirchhoff laws, nodal analysis, and modularity index. EVs offer high potentials to serve as mobile backup storage devices that can provide grid support to enhance its reliability as a means of smart grid application [14]. Reference [15] provides a Matlab-based Monte Carlo Simulation code that allows the incorporation of distributed energy resources (i.e., EVs) to assess the distribution network's reliability. Additionally, studies have covered the potentials of EVs in relevant frequency regulation and control. The authors of reference [16] proposed an intelligent aggregator that synchronizes the charging and discharging activities of a group of EVs in order to regulate frequency by compensating for any potential power deficiency. Similarly, the authors of reference [17] proposed a real-time dynamic decision-making framework based on Markov Decision Process (MDP) to allow intelligent frequency regulation by energy support from EVs. Reference [18] developed a multivariable generalized predictive controller to enable load frequency control in a standalone microgrid with V2G integration. The controller aims to allow sufficient energy exchange without causing frequency deficiency, considering possible load disturbances. Furthermore, the recent progressive policies that aim to reduce GHG emissions from the transportation sector will result in a mass acquisition of EVs in the next few years [19], especially in regions where utilization of EV is expected to have a significant reduction of GHG emissions as a result of their weather and energy grid mixes [20]. Such rapid adoption of EVs on a large scale without proper coordination can result in phase imbalance, equipment fallout and degradation, increase active and reactive power losses, among many problems [21]. Therefore, careful consideration needs to be given to overcome the problems that may arise due to the intermittency and stochastic nature of the energy sources on the hybrid microgrids.

In this work, an energy management and control strategy is developed to overcome the fluctuations of the voltage and frequency levels due to the presence of intermittent renewable energy sources and electric vehicle charging structure in hybrid microgrids. Furthermore, a hybridization algorithm of the Particle Swarm Optimization (PSO) and Applied Artificial Physics (APO) is utilized to tune the vector-decoupled control parameters of the interlinking converters to ameliorate the performance of the hybrid microgrid to achieve better resiliency and operation. Our proposed strategy highlights guaranteed stability of operation in the DC part of the microgrid while efficiently coordinate with the AC part during severe operating conditions such as high pulsed demands and islanding operation. The proposed algorithm is tested via a hardware-in-the-loop testbed at the Florida International University to for results verification. The results embolden the validity of our proposed strategy and hybridized algorithm to establish secure and safe active and reactive power exchange between the two sides of the hybrid microgrid without invoking an operational violation.

This paper is organized as follows: Section 2 shows in detail the system description and corresponding illustration of the modelling and control of the bidirectional converter, Section 3 presents the hybrid algorithm deployed in our work to provide the converter with optimized control parameters, Section 4 presents the experimental modelling and results of the proposed control mechanism, with Section 5 providing concluding remarks.

2. System's Description, Modeling, and Control

Figure 1 presents the system of study in this work which is implemented in hardware at the Energy Systems Research Laboratory group (ESRL) of the Florida International University (FIU). More information about the hardware testbed and its connections can be found in the previously published literature [22–24]. The hybrid microgrid at the testbed incorporates different harvested AC and DC sources that are integrated through interfaced power converters. Both sides are interlinked via a bidirectional converter, with the DC part contains PV systems, electric vehicles parking structure, and local and pulsed DC loads. On the other hand, the AC part of the microgrid is supplied with a synchronous generator as well as the typical load demands. During islanding operation, the microgrid is isolated and maintain its supply to local loads via both AC and DC sources. The microgrid is designed such that it can autonomously satisfy the energy demands without interruption under any circumstances. Two DC-to-DC boost converters are used in this work to link the DC components to the bidirectional power converter, as illustrated in Figure 1. Table 1 presents the parameters of the interlinking converter used in this work.

Table 1. Converter Parameters.

Parameter	Value
DC BUS Voltage	380 ± 20 V
Rating	10 kW
R_s,	0.01 ohm
C_{out}	1200 μF
Rc_{out}	0.008 ohm
L	12.7 mH
C_{in}	1200 μF
R_{cin}	0.008 ohm

where R_s, Rc_{out}, and R_{cin} are the resistance of the voltage source located at the DC side, resistances of the output and input capacitor of the power converter, while C_{out} and C_{in} are the values of the output and input capacitors.

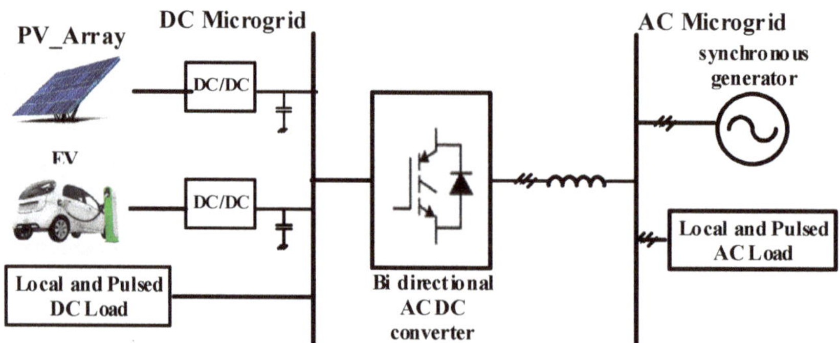

Figure 1. Schematic configuration of the hybrid microgrid in this work.

2.1. PV System Model and Interface

The PV system is modeled with a PV emulator that has a maximum capacity of 6 kW and can imitate a real-time PV system with different characteristics and under various operational conditions such as during temperature and irradiance changes. The PV emulator is constructed to utilize real-time algorithms that represent the PV array's mathematical models to generate reference power output from a programmable DC supply. Specifically, the PV models are established in Simulink and resembled in real-time operation via dSPACE following a graphic user interface (GUI). Accordingly, the emulator is then tested with real-time execution of the PV model considering various dynamic operational and steady-state conditions. Figure 2 presents the configuration of the laboratory PV emulator, which was first introduced by the authors in previous work [23]. Figure 3 shows a schematic illustration of the PV system connectivity with the boost converters for accurate integration with the microgrid's DC side. It is worth mentioning that the type of PV module in this work is the SPR-305-WHT PV system manufactured commercially by SunPower and can generate 305 watts as an output power per module with an efficiency of 18.9%. The current-voltage (IV)characteristic of the PV system could be represented by a single diode model as it provides accuracy and simplicity [25]. The current output of the PV arrays can be found by

$$I_{PV} = I_L - I_S \left[\exp\left(\frac{q(V_{PV} + I_{PV}R_s)}{K_B\,T\,A} - 1 \right) - \frac{q(V_{PV} + I_{PV}R_s)}{R_{sh}} \right] \tag{1}$$

where I_{PV} is the output current of the PV array, V_{PV} is the voltage reference which is established based on the perturbation and observation (P&O) algorithm that takes into consideration the temperature and solar irradiation levels, I_L is the internal PV current, I_S is the diode's reverse saturation current, R_{sh} is the parallel leakage resistance, R_s is the series resistance, A represents the ideality factor of the solar cell, while q is the charge of the electron that is assumed to be 1.6×10^{-19} C, and KB is the Boltzmann constant which is $1.3806488 \times 10^{-23}$ J/K. The DC-to-DC boost converter is used to step up the voltage level of the PV arrays to the voltage level of the microgrid's DC side when needed while ensure maximum power extraction based on the concept of maximum power point tracking (MPPT). Specifically, the P&O algorithm [26,27] is utilized in this work. This algorithm mainly depends on perturbing the voltage level of the PV panels by small magnitude (ΔV) and accordingly observing the change of power level (ΔP) to optimize the tracking of maximum power transfer from the arrays to take into consideration potential temperature variability during the day.

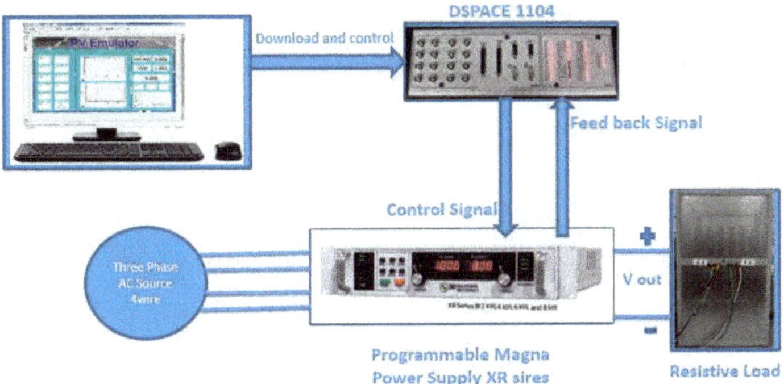

Figure 2. The PV emulator setup in our hardware testbed [23].

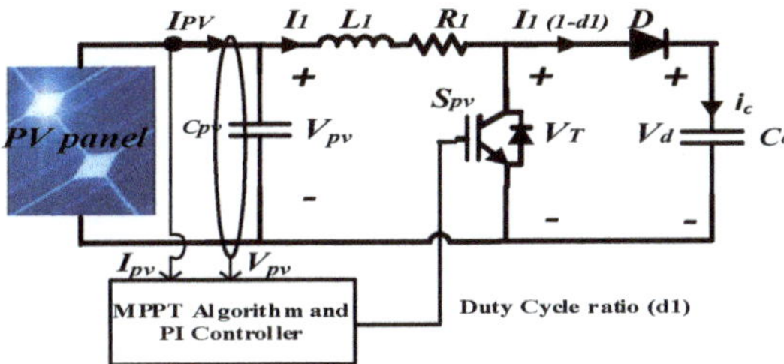

Figure 3. Schematic configuration of the PV system interface with the DC side of the hybrid microgrid.

The main contribution from utilizing the (P&O) algorithm in our work is to aim for zero difference of power received from the PV arrays in two successive iterations, denoted ΔP. This is accomplished by measuring the level of power at each iteration based on the PV output current so that the power level at the kth iteration is recorded, P_k, and $P_{(k+1)}$ for the following iteration. The DC-to-DC boost converter adjusts the power output of the PV system by either decreasing in the case of a negative ΔP or increasing it for the case of a positive ΔP. Once ΔP approaches zero, the PV system is said to be reaching its maximum power point (MPP). The process is updated iteratively throughout the operation hours to ensure maximum power production from the PV arrays. Figure 4 illustrates the described iteration process, while Figure 5 shows the implementation of the control process for the DC-to-DC converter in our work.

The control mechanism for the DC-to-DC boost converter is achieved based on the following formulas

$$V_{PV} - V_T = L_1 \frac{\partial L_1}{\partial t} + L_1 R_1 \tag{2}$$

$$I_{PV} - I_1 = C_{PV} \frac{\partial V_{PV}}{\partial t} \tag{3}$$

$$V_T = V_D(1 - D_1) \tag{4}$$

where L_1 and R_1 are bidrectional converter inductance and resistance, I_1 is the current corresponding to the duty-cycle ratio D_1 of the switch S_{PV}. V_T is the voltage across the switch, while V_{pv} is the voltage reference across capacitance C_{PV} and is determined following the utilization of the P&O algorithm based on the temperature and solar irradiation levels of PV arrays, as mentioned earlier. It should be noted that the control process is based on the dual-loop control mechanism. The inner current loop assists in the improvisation of the dynamic response, while the outer voltage loop keep tracks with the reference voltage levels given zero steady-state error.

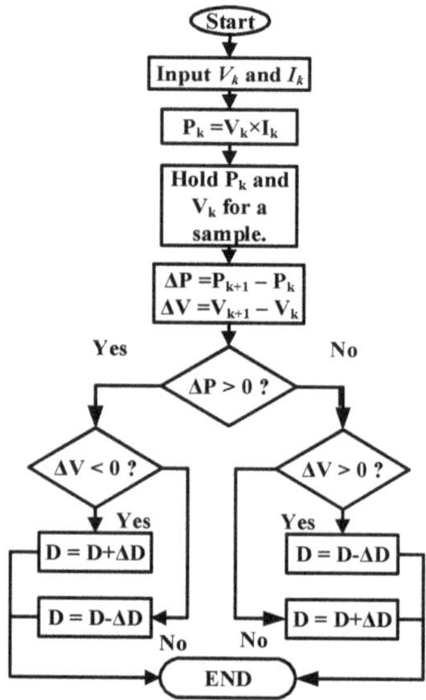

Figure 4. The perturbation and observation (P&O) algorithm utilized in this work.

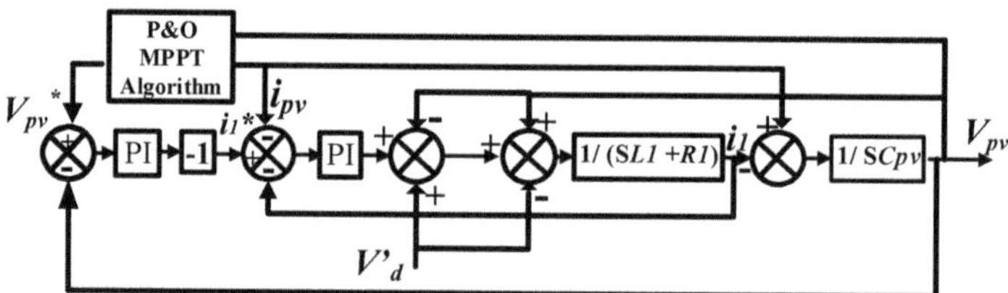

Figure 5. Block diagram for the boost converter control.

2.2. Electric Vehicles (EVs) Battery Converter Model and Control

The developed scheme for the electric vehicle charging converter is shown in Figure 6. The converter topology is a composed of a bidirectional DC-to-DC converter, with the EV is connected to the low-voltage side of the converter. The high-voltage side of the converter is directly connected with the microgrid DC bus. The converter is composed of two switches, Sc and Sd, each with its own operation mode and time. Specifically, switch Sc is on when the converter operates at the buck mode for charging activity during power transfer from the DC Bus to the EV's battery.

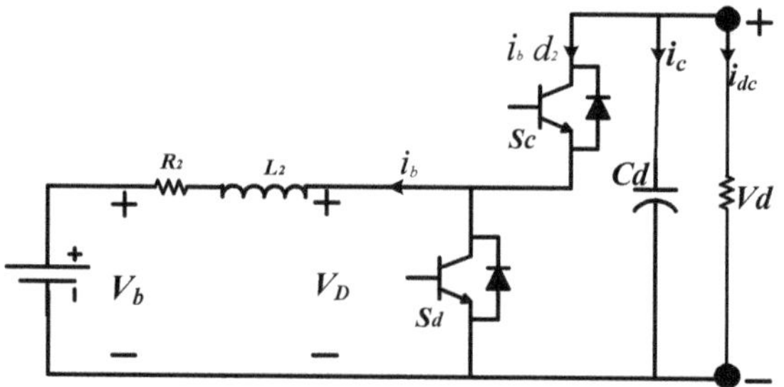

Figure 6. Configuration of the DC-to-DC converter interface with the electric vehicles (EVs) battery and the DC bus.

Conversely, switch Sd is on when the converter operates at the boost mode during the discharging process of the EV during power transfer from the EV's battery back to the DC bus. The control mechanism for the EV's battery converter is shown in Figure 6, and is mathematically illustrated as follows

$$V_D - V_b = L_2 \frac{\partial I_b}{\partial t} + R_2 I_b \tag{5}$$

$$V_D = V_D D_2 \tag{6}$$

$$I_1(1 - D_1) - i_{ac} - i_{dc} - I_b D_2 = C_D \frac{\partial V_D}{\partial t} \tag{7}$$

where V_d and i_{dc} are the DC side's voltage and current, i_{ac} is the current corresponding to the AC side of the hybrid microgrid, V_b and I_b is the EV battery's voltage and current, C_D is the bidirectional converter capacitance for boost mode, L_2 and R_2 are bidirectional converter inductance and resistance. It should be noted that the main task of the bidirectional DC-DC converter connected to the EVs charging structure is to regulate the DC bus voltage. To achieve this purpose, a dual-loop control is utilized to assist in providing a stable DC link voltage. Specifically, the external voltage-controlled loop establishes a reference charging current as a signal for the internal current-controlled loop. The difference between the reference and measured bus voltage serves as an input signal to the PI controller. This difference is used to measure the reference charging current since the internal current-controlled loop compares this estimated current-signal with that one of the referenced current flowing inside the converter. The produced output of this loop control serves as an input signal for a second PI controller for further optimization of the inner controller.

The EVs battery current i_b is calculated based on Equation (5), while the duty cycle is estimated by Equations (6) and (7). The discharging current can be calculated as follows

$$I_b = I_1(1 - D_1) - i_{ac} - i_{dc} \tag{8}$$

That is to say, if the voltage of the microgrid's DC bus is higher than the desired reference voltage signal, then the outer voltage control generates a negative reference current signal for the inner current-controlled loop. The generated signal current is used to adjust the correspondent duty cycle in order to influence the converter to operate in a buck mode only and suspend any discharging activity at the moment. On the other hand, if the voltage of the microgrid's DC bus is lower than the desired reference voltage signal, then the outer voltage control generates a positive reference current signal to regulate the current flow during the discharging process. As a result, an additional amount of energy is incurred and injected to the DC bus while improving the voltage profile at the moment. It is worth mentioning that the physical reference for the voltage source of the DC bus is 400 V.

2.3. Bidirectional DC-to-AC Converter Model and Control

In hybrid microgrids, managing robust frequency and voltage levels is challenging, especially during forced islanding operation where the AC side loses its connection to the grid's main slack bus. Typically, a hybrid microgrid owns synchronous generators that can manage load variations and maintain energy supply, even during islanded operation. However, high demands connected to the hybrid microgrid may lead to severe consequences such as frequency deficiency and potential voltage collapse. As a result, the bidirectional DC-AC converter's main task is to enable strict frequency and voltage regulation considering severe operational scenarios [28]. We consider this controller type to ensure a smooth power exchange between the DC and AC sides of the microgrid. The mathematical representation of the DC-AC converter model is illustrated as follows:

$$L_3 \frac{d}{dt} \begin{bmatrix} ia \\ ib \\ ic \end{bmatrix} + R_3 \begin{bmatrix} ia \\ ib \\ ic \end{bmatrix} = \begin{bmatrix} V_a \\ V_b \\ V_c \end{bmatrix} - \begin{bmatrix} e_a \\ e_b \\ e_c \end{bmatrix} + \begin{bmatrix} \Delta_a \\ \Delta_b \\ \Delta_c \end{bmatrix} \tag{9}$$

Considering D-Q coordinates, Equation (9) could be rewritten as follows

$$L_3 \frac{d}{dt} \begin{bmatrix} i_d \\ i_q \end{bmatrix} = \begin{bmatrix} -R_3 & wL_3 \\ -wL_3 & -R_3 \end{bmatrix} \begin{bmatrix} i_d \\ i_q \end{bmatrix} + \begin{bmatrix} V_{cd} \\ V_{cq} \end{bmatrix} - \begin{bmatrix} V_{sd} \\ V_{sd} \end{bmatrix} \tag{10}$$

The control mechanism of the bidirectional DC-AC converter utilized in our work is shown in Figure 7, where two-loop controllers, for both real and reactive power, are applied. The overall goal is to allow proper and intelligent control of both frequency and voltage levels at the hybrid microgrid. For frequency control, the difference between the measured frequency signals from that of the obtained reference frequency is established. The result is subtracted from the difference error between the measured and referenced DC voltage level, as described in part B of this section. The obtained value serves as an input signal to the PI controller, which initiates the current reference value, Id. Likewise, another control loop is deployed to achieve voltage stability employing optimized reactive power flow in the hybrid microgrid. This is made in the same manner as frequency control, where the difference between the measured and referenced voltage levels is calculated to produce a signal that serves as input to another PI controller to generate the Iq reference current. In the next section, we present a metaheuristic methodology based on the hybridization of Particle Swarm Optimization (PSO) and Artificial Applied Physics (APO) to tune the vector-decoupled control parameters illustrated in Figure 7 optimally.

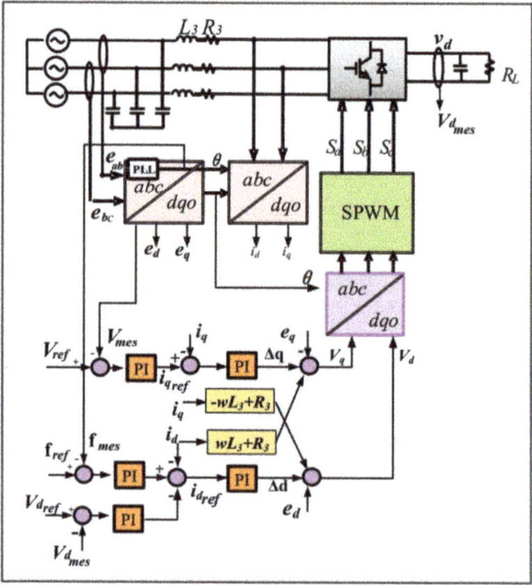

Figure 7. Schematic diagram of the bidirectional DC-AC converter.

3. Control Parameters Design Using Hybrid Artificial Physics Optimization-Particle Swarm Optimization (APOPSO) Algorithm

The central concept of applying our hybridization of PSO and APO is to integrate their individual strengths to establish an optimization algorithm that exhibits both the dominant global search abilities of the APO and efficient local exploration performance of the PSO while enhancing its convergence rate. In this work, the hybrid algorithm is utilized to optimize the vector-decoupled control parameters of the bidirectional converter to ensure efficient energy management driven by optimized variables while reducing the trial–error method described in Section 2. In a previous study, the authors developed an optimal reactive power dispatch study based on the hybrid APOPSO [29].

3.1. Artificial Physics Optimization (APO)

APO is a physics-based metaheuristic technique that is based on the idea of a gravitational metaphor that enables forces to produce attractiveness or repulsiveness movements on the articles that resemble the solutions of the optimization problem [29–31]. These movements represent the searching criteria for estimating the values of local and global optima. Furthermore, this is accomplished since the APO treats the examined parameters as physical objects that exhibit a mass with relative position and velocity. The mathematical description of the APO is as follows

$$m_i = g\left[f\left(x_i\right)\right] \tag{11}$$

When $f(x) \in [-\infty, \infty]$, then; arctan $[-f(x_1)] \in [\frac{-x}{2}, \frac{x}{2}]$, and $\tanh[-f(x_i)] \in [-I, I]$ with

$$\tanh(x_i) = \frac{e^x - e^{-x}}{e^x + e^{-x}} \tag{12}$$

where Equations (11) and (12) is mapped into the interval (0,1) via basic transformation function. Therefore, the mass functions of the APO is described as follows

$$m_i = e^{\frac{g\left[f(x_{best}) - f\left(x_i\right)\right]}{f(x_{worst}) - f(x_{best})}} \tag{13}$$

where $f\left(x_{best}\right)$ is the objective function corresponding to the position of the best-achieved value for the individual solution, which in this work resembles the best-obtained control parameter. On the other hand, $f\left(x_{worst}\right)$ refers to the value of the worst particular solution reported during the searching process. Both are represented as follows:

$$Best = avg\left\{\min f(x_i), \ i \in S\right\} \tag{14}$$

$$Worst = avg\left\{\max f\left(x_i\right), \ i \in S\right\} \tag{15}$$

where S is a set that is composed of N population of controlling parameters. A velocity vector is produced once each particle's mass is identified, with the level of exerted force influencing the change in velocities in an iterative manner. The amount of exerted force on each particle i (solution) can be found as follows

$$F_{ij,k} = \begin{cases} sgn\left(r_{ij}, k\right) \cdot G\left(r_{ij}, k\right) \cdot \frac{m_i m_j}{r_{ij}, k^2}; \ if \ f\left(x_j\right) < f(x_i) \\ sgn\left(r_{ji}, k\right) \cdot G\left(r_{ji}, k\right) \cdot \frac{m_i m_j}{r_{ij}, k^2}; \ if \ f\left(x_j\right) \geq f(x_i) \end{cases} \tag{16}$$

and

$$r_{ij, k} = x_{j,k} - x_{i,k} \tag{17}$$

where $F_{ij,k}$ is the kth force exerted on particle i via another particle j in their corresponding dimensions; $x_{i,k}$ and $x_{j,k}$ are the kth dimensional coordinates of the swarm particles i and j; $r_{ij,k}$ is the distance between the two measured coordinates. $Sgn(r)$ represents the signum function, whereas $G(r)$ depicts the gravitational factor that follows the changes on $r_{ij,k}$ iteratively, both represented mathematically as follows

$$Sgn(r) = \begin{cases} 1 & if & r \geq 0 \\ -1 & if & r < 0 \end{cases} \tag{18}$$

$$G\left(r\right) = \begin{cases} g|r|^h & if & r \leq 1 \\ g|r|^q & if & r > 1 \end{cases} \tag{19}$$

In a thorough manner, the total force applied on all particles (control parameters of study) can be modeled as:

$$F_{i,k} = \sum_{\substack{j=1 \\ i \neq j}}^{m} F_{ij,\,k} \qquad \forall\, i \neq best \tag{20}$$

One crucial aspect to consider when deploying the APO to solve an optimization problem is the understanding of its particles' motion paradigm in the solution space. Specifically, the measured force could be used to estimate the velocity of the moving particles and therefore find in an iterative fashion their respected positions in the solution space. Such motion paradigm is set in a two- or three-dimensional space and is modeled as follows:

$$V_{i,k}\,(z+1) = w \cdot V_{i,k}(t) + \beta * \frac{F_{i,k}}{m_i} \tag{21}$$

$$x_{i,k}\,(t+1) = X_{i,k}(t) + V_{i,k}(t+1) \tag{22}$$

$V_{i,k}$ and $x_{i,k}$ represent the kth velocity and distance components corresponding to particle i during an iteration t, while β is a uniformly distributed random variable within the interval $[0,\,1]$ and w is a user-defined inertia weight that is updated iteratively and usually assume a value between the interval 0.1 to 0.99. Furthermore, the inertia weight is a good indication of the level of performance of the APO algorithm, with higher values of w indicates greater velocity changes. It should be noted that at each iteration, each particle identifies the information of its nearby particles (solutions) which emphasizes the great search strategy of the APO. Once an iteration is performed, all the particles' relative positions are identified and consequently the objective fitness function adjusts to the newly obtained positions. A stopping criterion is enabled once a pre-determined number of iterations are reached without a significant difference in the obtained best particle position.

3.2. The Particle Swarm Optimization (PSO)

Considered one of the most popular metaheuristic techniques, PSO is a bio-inspired, population-driven algorithm first presented by Kennedy in [32]. PSO advances based on evolutionary computations with a sample of preliminary randomized solutions at the first iteration, updated iteratively to establish local and global optima values. The obtained solutions are deemed particles that fly in the solution space with a determined velocity from preceding iterations. It should be noted that the obtained velocity and position values of each solution set are updated iteratively as follows:

$$V_{ij}(t+1) = [W * V_{ij}(t)] + [C_1 + r_1 + [Pbest_{ij} - X_{ij}(t)]] + [C_2 + r_1 + [gbest_{ij} - X_{ij}(t)]] \tag{23}$$

$$X_{ij}(t+1) = X_{ij}(t) + C\, V_{ij}(t+1) \tag{24}$$

$X_{ij}(t)$ and $V_{ij}\,(t)$ are both vector representations of velocity and position in the solution space for particle i, whereas *Pbest* and *gbest* stand for the best individual and global obtained solutions, respectively. The popularity of PSO as a well-established and referred metaheuristic algorithm is attributed to its efficient searching strategy along with prematurely convergence rates without the requisite of finding a local optimum in first place.

3.3. The Hybridization of APO and PSO to Optimize the Vector-Decoupled Control Parameters

The hybridization of APO and PSO is to establish in this work to take advantage of their individual strengths that lead to an overall improvement in the optimization process. Specifically, such integration utilizes the high efficiency of global search of the APO with the strong local exploratory search of the PSO while significantly enhancing its convergence rate. In this paper, the two algorithms are integrated following a low-level heterogeneous routine. As a consequence, the velocity and positions equations are modified as follows:

$$v_{i,k}(t+1) = W \cdot v_{i,k}(t) + \beta_1 - r_1 \cdot \left[\frac{F_{i,k}(t)}{m_i}\right] + \beta_2 \cdot r_2 \cdot \left[g\,best - x_{i,k}(t)\right] \tag{25}$$

$$x_{i,k}(t+1) = x_{i,k}(t) + v_{i,k}(t+1) \tag{26}$$

This proposed hybridization allows parallel search within a set of population which leads to avoidance of getting trapped in local optima. The control parameters to be optimized are K_{p_f}, K_{i_f}, K_{p_vdc}, K_{i_vac}, K_{p_m}, K_{i_vdc}, K_{i_m}, and K_{p_vac}. Following the microgrid's dynamic simulation, the hybrid APOPSO algorithm evaluates the integral absolute values of both the frequency (Δf) and the RMS voltage (ΔV_{rms}) deviation levels corresponding to the AC part of the microgrid. In this paper, two fitness functions are applied to the hybrid algorithm to properly estimate the self-tuning of the gains of the PI controller. The output of the fitness functions is used to control the power-sharing levels of the bidirectional converter, as follows

$$min\{F = e(t) = y(t) * -y(t)\} \tag{27}$$

$$MOF = Min\left\{\int_{t_0}^{t_f} x_f\,|\Delta f|dt + x_v \int_{t_0}^{t_f} |\Delta V_{rms}|dt\right\} \tag{28}$$

where *e(t)* represents the level of the errors, *y(t)** the desired value to be obtained and *y(t)* is the actual measured value per each iteration, while x_f and x_v represent penalty factors to enforce the voltage and frequency levels to be within the desired limits. Figure 8 shows the flowchart of the proposed hybrid algorithm to optimize our control parameters.

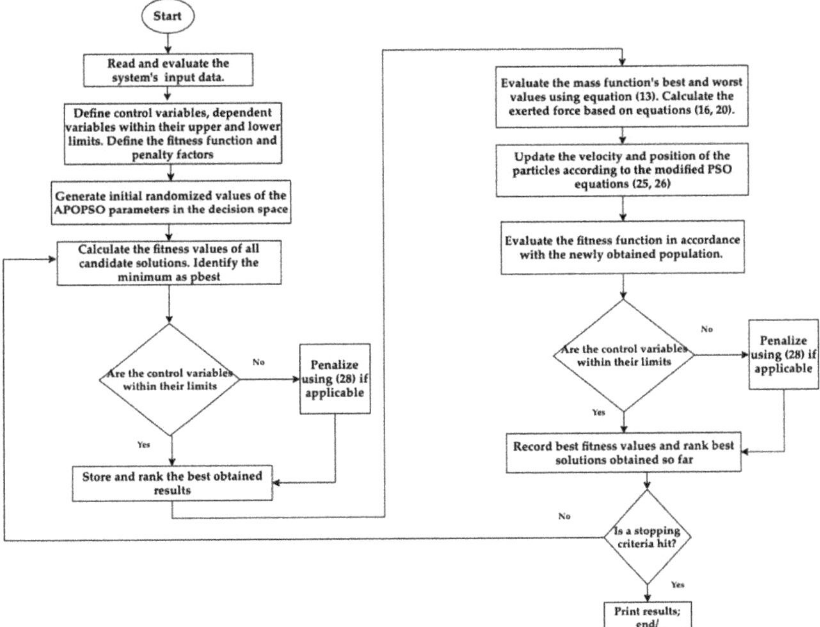

Figure 8. The proposed hybrid APOPSO applied to the energy management and control of hybrid microgrids.

The purpose of utilizing this search strategy is to ensure safe and optimal sharing of the power between the two sides of the hybrid microgrid in terms of providing the bidirectional controller with optimized vector-decoupled control parameters to achieve ideal converter's operation and ensure operation within system's limits which are defined in our study not to exceed ±5% of the frequency, and ±8% of the base voltage levels. The results of applying this hybrid algorithm is shown in Table 2 with the produced optimal parameters to ensure optimized damping performance. Figure 9 presents the convergence performance of the proposed algorithm while Figure 10 shows the results for best individual results per each of the eight variables in our study.

The searching criteria stop if any of the following conditions have been reached; (i) The hybrid algorithm reached the maximum allowed number of iterations, (ii) Same solutions have been obtained for a predetermined number of iterations, or (iii) Same set of solutions (by means of particles) are found in the same solution space.

Table 2. The optimal control parameters.

Variable	Obtained Optimal Value
K_{p_f}	1.769
$K_{i\_f,}$	2856.447
$K_{p\_vdc,}$	0.604
$K_{i\_vdc,}$	1220.302
K_{p_vac}	0.025
K_{i_vac}	0.079
K_{p_m}	556.071
K_{i_m}	4810.291

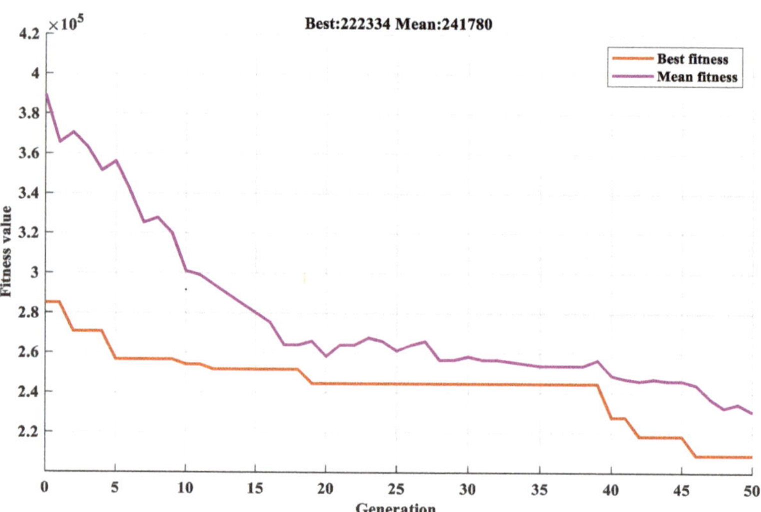

Figure 9. Convergence performance of the proposed algorithm.

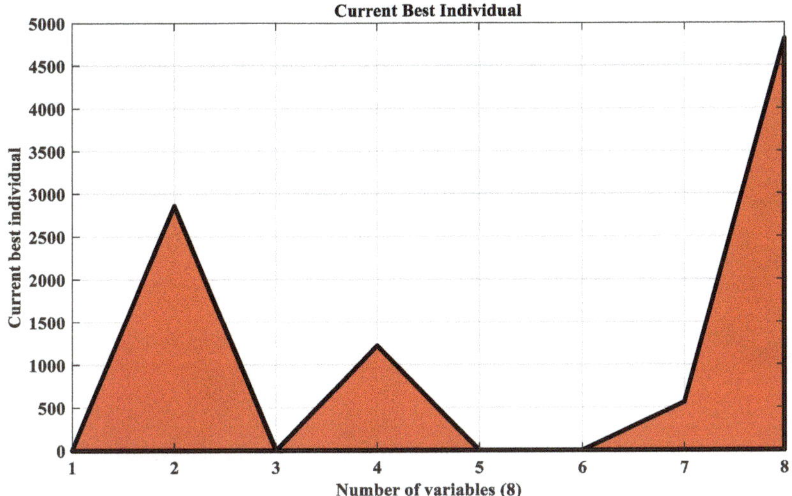

Figure 10. The optimization results of the proposed hybrid algorithm.

4. Experimental Results

To verify our proposed methodology, we demonstrate its concept via hardware-in-the-loop testbed at the Florida International University. A MATLAB/Simulink model is built and is shown in Figure 11. Specifically, it resembles the hybrid microgrid that consists of a synchronous generator and programmable AC loads connected at the AC side, with a PV emulator and a lithium–ion battery to resemble EVs activities at the DC side along with programmable DC loads. An interlinking bidirectional converter is utilized to connect the AC/DC sides of the microgrid. Figure 12 shows the hardware components at our testbed lab to perform this study. We set the simulation time to be 4 seconds and applied the vector-decoupled with the optimization parameters obtained via the hybrid algorithm, as illustrated in this work. Figures 13–15 show the results, with the output of the PV system dropping from 0.8 to 1.5 s as a result of a hypothetical cloud-dense during a specific time of the day, as shown in Figure 13a.

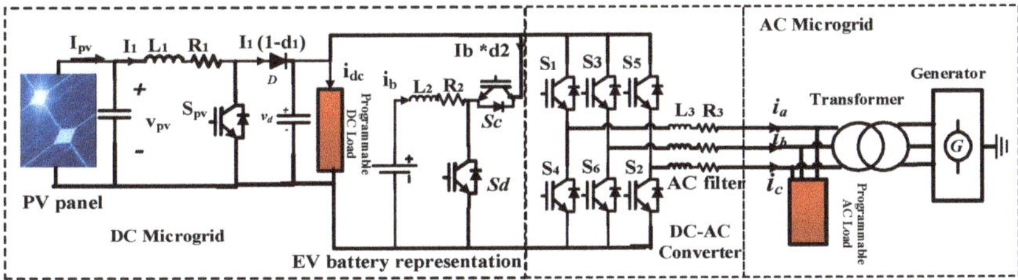

Figure 11. Schematic illustration of the hybrid microgrid connection at our testbed.

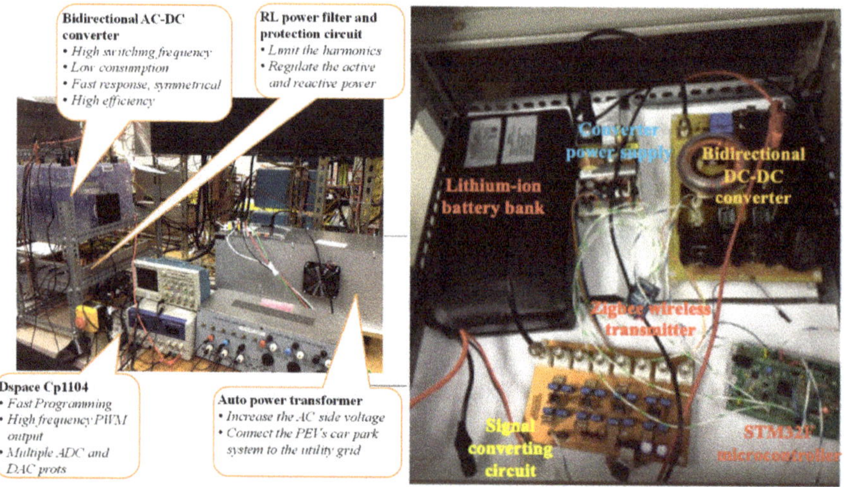

Figure 12. Hardware-in-the-loop equipment at our testbed.

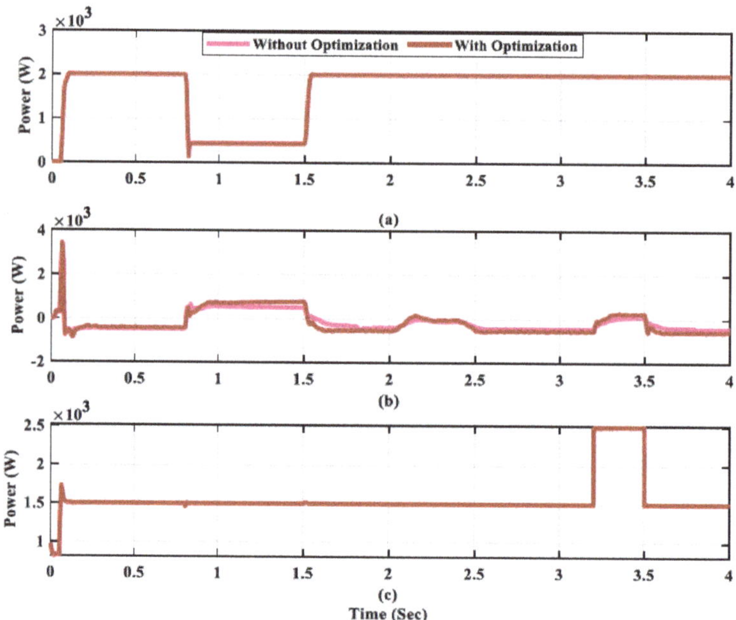

Figure 13. Results of the DC side: (**a**) Power generation from the PV system (**b**) EV's battery power (**c**) Load power.

To be able to manage such deficiencies in PV generation, the microgrid operator allows more EVs discharging events via reduced monetary incentives to encourage the consumers to discharge during once such situation incur at any potential time of the day, as shown in Figure 13b. In order to compensate for any potential lack of discharging due to the randomness of consumer participation, the AC generator increases its output to meet the remaining loads to keep the microgrid's operation in balance. This is demonstrated in Figure 15a and is achieved in a rapid manner to keep the system's voltage and frequency levels unaffected. Figures 13c and 15c present the load profiles, where considering the optimization of the vector-decoupled parameters, based on our hybrid algorithm, lead to more contribution from the AC generator side. It noted that assuming coordinated large-scale participation of EVs discharging, the stress on the synchronous generator could be furtherly alleviated. Such incorporation of EVs in the balancing criterion could be then estimated, at the discretion of the microgrid's tertiary control, which eventually contribute to a smarter charging and discharging scheduling.

As noted from Figure 14a–c, the pulsed load of the DC side is energized for a total duration of 0.4 seconds between the timeslots 2 to 2.4 s, with another energized pulsed load between the timeslot 3.2 to 3.5 s. Following our proposed mechanism, the controller performs the controlling procedure accordingly and mitigates the pulsed loads by balancing the power-sharing to a proper ratio to prevent any potential disturbances on the microgrid operation. Specifically, the controller force reversed power flow to the DC part of the hybrid microgrid if the DC loads are energized. Such a process is shown in Figure 15b in the case of negative power, which is an indication of power flow from the AC part of the grid to its DC side to compensate for the deficiency at the DC voltage level. As shown in Figure 14, our controlling mechanism achieves stable and secure microgrid operation by acceptable variations of the frequency and voltage levels. Although voltage variations are a bit high, we emphasize that they remain within a safe and acceptable level of operation. Figure 15c shows the DC side voltage level and is stable around the reference value of 400 V. As expected, variations of the generator output lead to fluctuations of frequency levels that exceed allowable and safe limits, which could trigger the operation of under- or over-frequency protection relays. However, these fluctuations are significantly reduced and managed following our proposed control mechanism based on optimized parameters using APOPSO. This shows the robustness and effectiveness of our proposed technique.

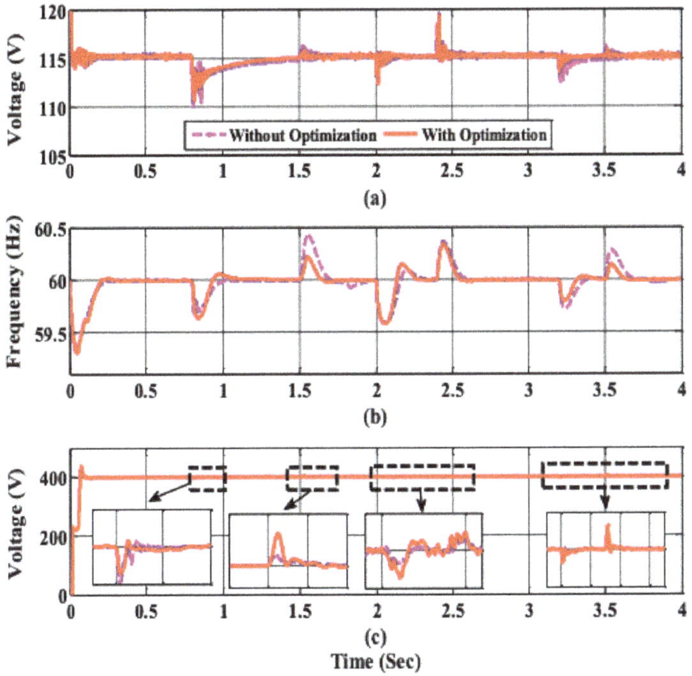

Figure 14. (**a**) AC voltage (RMS value) of phase a (**b**) AC side frequency level (**c**) DC voltage level.

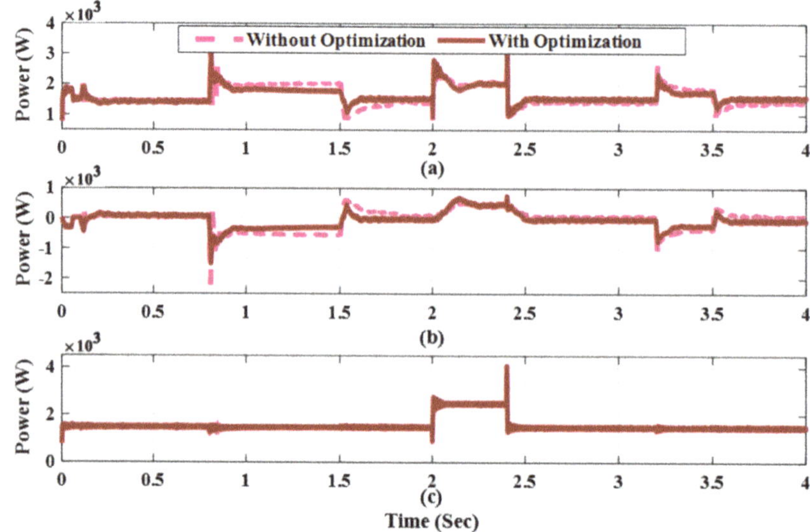

Figure 15. Results of the AC side of the hybrid microgrid: (**a**) AC generator output power (**b**) Inverter power at the point of common coupling (**c**) Load power.

5. Conclusions

In this work, a metaheuristic-based vector-decoupled algorithm for hybrid microgrid energy control and management is proposed. The algorithm aims to ensure safe and stable power-sharing between the DC and AC parts of the microgrid considering variable renewable energy sources, EV charging structure, as well as severe operational condition such as in the case of forced islanding operation. The metaheuristic algorithm provides the interlinking converter with optimized parameters to manage the microgrid's operation under various load and resources conditions. mechanism enables a smart and rapid. A hardware-in-the-loop implementation verifies and validates the proposed technique and offer stable and robust operation even during islanding situation. Furthermore, stable voltage and frequency levels are achieved and the power sharing between the two parts of the microgrid is accomplished. Specifically, we assumed a reduction at the power level of the DC side due to dense cloud in the time between 0.8s to 1.5s, as shown in Figure 13a. Accordingly, the controller requests more energy discharge from the EVs during this period to compensate for this deficiency, as illustrated in Figure 13b, while it allows for power sharing from the synchronous generator located at the AC side as shown in Figure 15a to assist the deficiency in the DC side. This is pivotal in the balancing of the operation especially in the case of insufficient participation of the EVs to discharge their energy during the scenario of reduced PV output. It is noted from the results that this has been achieved in rapid and robust manner without impacting the load levels. It should be noted that the parameter optimization of the proposed hybrid algorithm allows more participation from the AC side. Since large variations in the generator output may lead to frequency fluctuations, optimization of the parameters is required in this work. This is achieved by optimizing those parameters using the proposed APOPSO algorithm. As can be shown in Figure 14c, the optimized parameters reduced the fluctuations significantly in comparison with case of non-optimized parameters. Fluctuations in the non-optimization scenario may harm the operation of the hybrid microgrids and could trigger false operation of the over/under frequency protection relays. The success of the hybrid algorithm in reducing the fluctuations indicate its robustness and effectiveness in the hybrid microgrid energy management and control.

Future work is expected to incorporate algorithms that propose dynamic pricing structure to accurately reflect the real-time energy prices as result of control activities in hybrid microgrids. Soon, huge participation of EVs as well as privately owned small-scale PV systems is expected, and a fair pricing structure will be required to encourage more participation from consumers sides. The authors of this work propose a new pricing scheme that allocates special pricing tariff on electric vehicles that charge considering stochastic microgrids operation and energy management [33]. Furthermore, the authors suggest that this area of research needs further investigation. Additionally, future work is also anticipated in regard with machine learning applications in smart control of power quality problems as a result of large adoption of EVs in hybrid microgrids. In such studies, smart control is integrated to enhance the voltage fluctuations and harmonics as result of stochastic large-scale integration of EVs activities on microgrids.

Author Contributions: Conceptualization, T.M.A. and A.F.E.; Methodology, T.M.A. and A.F.E. Software, T.M.A. and A.F.E.; Validation, T.M.A, A.F.E and O.M.; Formal Analysis, T.M.A.; Investigation, T.M.A.; Resources, T.M.A.; Data Curation, T.M.A. and A.F.E.; Writing-Original Draft Preparation, T.M.A.; Writing-Review and Editing, T.M.A, A.F.E and O.M.; Visualization, T.M.A. and A.F.E.; Supervision, O.M.; Project Administration, O.M.; Funding Acquisition, T.M.A. All authors have read and agreed to the published version of the manuscript.

Funding: We acknowledge the financial support of Taibah University, Saudi Aribia for author Tawfiq Aljohani. Laboratory funding support by the Energy Systems Research Laboratory, Florida International University, Miami, FL 33174 (email: mohammed@fiu.edu).

Conflicts of Interest: The authors declare no conflict of interest.

Nomenclature

EV:	Electric Vehicle
PSO:	Particle Swarm Optimization
APO:	Artificial Physics Optimization
APOPSO:	Hybridization of the PSO and APO
GHG:	Greenhouse Gasses
RES:	Renewable Energy Sources
GUI:	Graphic User Interface
IV:	Current-Voltage Characteristic Curve
MPPT:	Maximum Power Point Tracking
P&O:	Perturbation and Observation
I_L:	Internal PV Current
I_{pv}:	PV Array Output Current
I_S:	The Diode's Reverse Saturation Current
R_{sh}:	The Parallel Leakage Resistance
Rs:	The Series Resistance
q:	The Electron's Charge (1.6×10^{-19} C)
K_B:	Boltzmann Constant ($1.3806488 \times 10^{-23}$ J/K)
d_1:	Duty Cycle Ratio of the Converter
V_{pv}:	The Voltage Reference of the PV
V_D:	The Voltage of the DC Side of the Hybrid Microgrid
V_T :	The voltage level across the bidirectional switch.
i_b:	EVs Battery Current
C_D :	The bidirectional converter capacitance for boost mode
L, R:	The bidirectional converter resistance and inductance.
i_{dc}:	Current Correspondence to the DC Side
i_{ac}:	Current Correspondence to the AC Side
I_q:	The PI controller reference current
$F_{ij,k}$:	The kth force applied on particle i via another particle j
$x_{i,k}$, $x_{j,k}$:	The kth dimensional coordinates for swarm particles i and j
$r_{ij,\,k}$:	Distance between two coordinates
$G(r)$:	The gravitational factor

$V_{i,k}$:   The kth component of particle i 's velocity at iteration t
$x_{i,k}$:   The kth component of particle i 's distance at iteration t
Pbest: Best Local Solution for an Individual Swarm
gbest: Best Global Solution
x_f, x_v: Penalty factors to enforce the voltage and frequency levels

References

1.   Tran, M.; Banister, D.; Bishop, J.D.; McCulloch, M.D. Realizing the electric-vehicle revolution. *Nat. Clim. Chang.* **2012**, *2*, 328–333. [CrossRef]
2.   Harrison, R.M.; Hester, R.E.; Powlesland, C. *Environmental Impact of Power Generation*; Royal Society of Chemistry: London, UK, 2007.
3.   Wu, D.; Tang, F.; Dragicevic, T.; Vasquez, J.C.; Guerrero, J.M. A control architecture to coordinate renewable energy sources and energy storage systems in islanded microgrids. *IEEE Trans. Smart Grid* **2014**, *6*, 1156–1166. [CrossRef]
4.   Lasseter, R.H.; Paigi, P. Microgrid: A conceptual solution. In Proceedings of the 2004 IEEE 35th Annual Power Electronics Specialists Conference (IEEE Cat. No. 04CH37551), Aachen, Germany, 20–25 June 2004; Volume 6, pp. 4285–4290.
5.   Hatziargyriou, N.; Asano, H.; Iravani, R.; Marnay, C. Microgrids. *IEEE Power Energy Mag.* **2007**, *5*, 78–94. [CrossRef]
6.   Karavas, C.S.; Kyriakarakos, G.; Arvanitis, K.G.; Papadakis, G. A multi-agent decentralized energy management system based on distributed intelligence for the design and control of autonomous polygeneration microgrids. *Energy Convers. Manag.* **2015**, *103*, 166–179. [CrossRef]
7.   Karavas, C.S.; Arvanitis, K.; Papadakis, G. A game theory approach to multi-agent decentralized energy management of autonomous polygeneration microgrids. *Energies* **2017**, *10*, 1756. [CrossRef]
8.   Rocabert, J.; Luna, A.; Blaabjerg, F.; Rodriguez, P. Control of power converters in AC microgrids. *IEEE Trans. Power Electron.* **2012**, *27*, 4734–4749. [CrossRef]
9.   Farahat, M.A.; Metwally, H.M.B.; Mohamed, A.A.E. Optimal choice and design of different topologies of DC–DC converter used in PV systems, at different climatic conditions in Egypt. *Renew. Energy* **2012**, *43*, 393–402. [CrossRef]
10.   Lee, J.P.; Min, B.D.; Kim, T.J.; Yoo, D.W.; Yoo, J.Y. A novel topology for photovoltaic DC/DC full-bridge converter with flat efficiency under wide PV module voltage and load range. *IEEE Trans. Ind. Electron.* **2008**, *55*, 2655–2663. [CrossRef]
11.   Wang, T.; O'Neill, D.; Kamath, H. Dynamic control and optimization of distributed energy resources in a microgrid. *IEEE Trans. Smart Grid* **2015**, *6*, 2884–2894. [CrossRef]
12.   Karavas, C.S.; Arvanitis, K.G.; Kyriakarakos, G.; Piromalis, D.D.; Papadakis, G. A novel autonomous PV powered desalination system based on a DC microgrid concept incorporating short-term energy storage. *Sol. Energy* **2018**, *159*, 947–961. [CrossRef]
13.   Aljohani, T.M.; Saad, A.; Mohammed, O. On the Real-Time Modeling of Voltage Drop and Grid Congestion Due to the Presence of Electric Vehicles on Residential Feeders. In Proceedings of the 2020 IEEE International Conference on Environment and Electrical Engineering and 2020 IEEE Industrial and Commercial Power Systems Europe (EEEIC/I&CPS Europe), Madrid, Spain, 9–12 June 2020.
14.   Aljohani, T.M.; Beshir, M.J. Distribution system reliability analysis for smart grid applications. *Smart Grid Renew. Energy* **2017**, *8*, 240–251. [CrossRef]
15.   Aljohani, T.M.; Beshir, M.J. Matlab code to assess the reliability of the smart power distribution system using monte carlo simulation. *J. Power Energy Eng.* **2017**, *5*, 30–44. [CrossRef]
16.   Han, S.; Han, S.H.; Sezaki, K. (2010, January). Design of an optimal aggregator for vehicle-to-grid regulation service. In Proceedings of the 2010 Innovative Smart Grid Technologies (ISGT), Gothenburg, Sweden, 19–21 January 2010; pp. 1–8.
17.   Islam, M.M.; Zhong, X.; Sun, Z.; Xiong, H.; Hu, W. Real-time frequency regulation using aggregated electric vehicles in smart grid. *Comput. Ind. Eng.* **2019**, *134*, 11–26. [CrossRef]
18.   Yang, J.; Zeng, Z.; Tang, Y.; Yan, J.; He, H.; Wu, Y. Load frequency control in isolated micro-grids with electrical vehicles based on multivariable generalized predictive theory. *Energies* **2015**, *8*, 2145–2164. [CrossRef]

19. Ou, X.; Zhang, X.; Zhang, X.; Zhang, Q. Life cycle GHG of NG-based fuel and electric vehicle in China. *Energies* **2013**, *6*, 2644–2662. [CrossRef]
20. Aljohani, T.M.; Alzahrani, G. Life Cycle Assessment to Study the Impact of the Regional Grid Mix and Temperature Differences on the GHG Emissions of Battery Electric and Conventional Vehicles. In Proceedings of the 2019 SoutheastCon, Huntsville, AL, USA, 11–14 April 2019; pp. 1–9.
21. Shaaban, M.F.; Atwa, Y.M.; El-Saadany, E.F. PEVs modeling and impacts mitigation in distribution networks. *IEEE Trans. Power Syst.* **2012**, *28*, 1122–1131. [CrossRef]
22. Salehi, V.; Mohamed, A.; Mazloomzadeh, A.; Mohammed, O.A. Laboratory-based smart power system, part I: Design and system development. *IEEE Trans. Smart Grid* **2012**, *3*, 1394–1404. [CrossRef]
23. Ebrahim, A.F.; Ahmed, S.M.; Elmasry, S.E.; Mohammed, O.A. Implementation of a PV emulator using programmable DC power supply. In Proceedings of the SoutheastCon 2015, Fort Lauderdale, FL, USA, 9–12 April 2015; pp. 1–7.
24. Ghareeb, A.T.; Mohamed, A.A.; Mohammed, O.A. DC microgrids and distribution systems: An overview. In Proceedings of the 2013 IEEE Power & Energy Society General Meeting, Vancouver, BC, Canada, 21–25 July 2013; pp. 1–5.
25. Bacha, S.; Picault, D.; Burger, B.; Etxeberria-Otadui, I.; Martins, J. Photovoltaics in Microgrids: An Overview of Grid Integration and Energy Management Aspects. *IEEE Ind. Electron. Mag.* **2015**, *9*, 33–46. [CrossRef]
26. Hussein, K.H.; Muta, I.; Hoshino, T.; Osakada, M. Maximum photovoltaic power tracking: An algorithm for rapidly changing atmospheric conditions. *IEE Proc.-Gener. Transm. Distrib.* **1995**, *142*, 59–64. [CrossRef]
27. Liu, X.; Lopes, L.A. An improved perturbation and observation maximum power point tracking algorithm for PV arrays. In Proceedings of the 2004 IEEE 35th Annual Power Electronics Specialists Conference (IEEE Cat. No. 04CH37551), Aachen, Germany, 20–25 June 2004; Volume 3, pp. 2005–2010.
28. Elsayed, A.; Ebrahim, A.F.; Mohammed, H.; Mohammed, O.A. Design and implementation of AC/DC active power load emulator. In Proceedings of the SoutheastCon 2015, Fort Lauderdale, FL, USA, 9–12 April 2015; pp. 1–5.
29. Aljohani, T.M.; Ebrahim, A.F.; Mohammed, O. Single and multiobjective optimal reactive power dispatch based on hybrid artificial physics–particle swarm optimization. *Energies* **2019**, *12*, 2333. [CrossRef]
30. Spears, W.M.; Gordon, D.F. Using artificial physics to control agents. In Proceedings of the 1999 International Conference on Information Intelligence and Systems (Cat. No. PR00446), Bethesda, MD, USA, 31 October–3 November 1999; pp. 281–288.
31. Xie, L.; Zeng, J.; Cui, Z. General framework of artificial physics optimization algorithm. In Proceedings of the 2009 World Congress on Nature & Biologically Inspired Computing (NaBIC), Coimbatore, India, 9–11 December 2009; pp. 1321–1326.
32. Kennedy, J.; Eberhart, R. Particle swarm optimization. In Proceedings of the ICNN'95-International Conference on Neural Networks, Perth, Australia; 2009; Volume 4, pp. 1942–1948.
33. Aljohani, T.M.; Ebrahim, A.; Mohammed, O. Dynamic Real-Time Pricing Structure for Electric Vehicle Charging Considering Stochastic Microgrids Energy Management System. In Proceedings of the 2020 IEEE International Conference on Environment and Electrical Engineering and 2020 IEEE Industrial and Commercial Power Systems Europe (EEEIC/I&CPS Europe), Madrid, Spain, 9–12 June 2020.

Article

Real-Time Building Smart Charging System Based on PV Forecast and Li-Ion Battery Degradation

Wiljan Vermeer *, Gautham Ram Chandra Mouli and Pavol Bauer

Electrical Sustainable Energy Department, Faculty of Electrical Engineering, Mathematics and Computer Science, Delft University of Technology, 2628CD Delft, The Netherlands; g.r.chandramouli@tudelft.nl (G.R.C.M.); p.bauer@tudelft.nl (P.B.)
* Correspondence: w.w.m.vermeer@tudelft.nl

Received: 20 May 2020; Accepted: 28 June 2020; Published: 2 July 2020

Abstract: This paper proposes a two-stage smart charging algorithm for future buildings equipped with an electric vehicle, battery energy storage, solar panels, and a heat pump. The first stage is a non-linear programming model that optimizes the charging of electric vehicles and battery energy storage based on a prediction of photovoltaïc (PV) power, building demand, electricity, and frequency regulation prices. Additionally, a Li-ion degradation model is used to assess the operational costs of the electric vehicle (EV) and battery. The second stage is a real-time control scheme that controls charging within the optimization time steps. Finally, both stages are incorporated in a moving horizon control framework, which is used to minimize and compensate for forecasting errors. It will be shown that the real-time control scheme has a significant influence on the obtained cost reduction. Furthermore, it will be shown that the degradation of an electric vehicle and battery energy storage system are non-negligible parts of the total cost of energy. However, despite relatively high operational costs, V2G can still be cost-effective when controlled optimally. The proposed solution decreases the total cost of energy with 98.6% compared to an uncontrolled case. Additionally, the financial benefits of vehicle-to-grid and operating as primary frequency regulation reserve are assessed.

Keywords: smart charging; electric vehicle; vehicle to grid; V2G; battery degradation; Li-ion; real-time; moving horizon window

---

1. Introduction

In 2015, transportation accounted for 19% of global energy consumption, almost all of which was powered by fossil fuels (including electric vehicles (EVs) and plug-in hybrid EVs) [1]. Fortunately, the cost of EVs is drastically reducing and their market share is increasing. However, for EVs to be truly sustainable, they have to be charged from a sustainable energy source. Photovoltaïc (PV) solar energy is now being investigated as a primary energy source for EV charging due to the synergies which exist between EV and PV. As both are inherently DC, directly charging an EV from PV power increases charging efficiency and charger density. Furthermore, an EV in combination with vehicle-to-grid (V2G) can act as a storage, can reduce the intermittent character of PV, can provide ancillary services, and can act as a primary energy source for other loads [2,3]. Finally, charging an EV from local PV power reduces the stress which EV charging is imposing on the future grid.

Another significant part of global energy consumption is the built environment; In [4,5] it is stated that the built environment emits up to 40% of all global greenhouse gasses. In the future, the phasing out of natural gas will increase the electrical demand of buildings as heat pumps (HPs) will be used for building heating. However, often, the existing distribution grid is unable to provide this increase in electrical demand caused by HPs and EVs. Luckily, Battery Energy Storage (BES) systems, EV/V2G, and locally produced PV power can help in providing this power and therefore can reduce the grid stresses while at the same time increase the renewable energy consumption. However, getting the most out of these mutual benefits requires complex charging algorithms based on load and PV power forecasts.

In this paper, a real-time building smart charging algorithm is presented, which, based on forecasts and Li-ion battery degradation, minimizes the operational costs of a PV-EV-BES-HP system while at the same providing a supporting role in the future smart grid by ancillary services and demand-side management.

1.1. Literature Study

The most straightforward control scheme of any EV/BES-PV system is to use a rule-based control scheme, where the current state of the system determines the next action, such as that presented in [6,7]. However, the effectiveness of rule-based schemes is limited, as future supply or demand is not anticipated and their operation is not close to optimal. Therefore, these systems are not investigated further. In [8–13], residential building-based smart charging systems are presented in which the energy costs are minimized. In [8,9], a time-series model is used to predict PV power and residential electrical demand; however, a coarse resolution of 1 h is used, which can lead to significant forecasting errors. In [9,10], also thermal storage and shiftable appliances are taken into account. In this study, these are considered non-flexible due to the low amount of flexibility which can be obtained and the high amount of comfort which is compromised. In [14], a mixed-integer linear programming problem that minimizes the charging costs with 30 min time steps is presented. However, forecasts or battery degradation costs are not taken into account. A hierarchical distributed smart charging station is proposed in [15]. Here, the individual systems try to stabilize their average available capacity of the battery storage bank, while the objective of a single EV is to maximize their charging power. Also, here, no regard for forecasting or degradation is taken into account. In [16], first, a two-stage optimization problem day-ahead scheduling is performed based on stochastic programming. Next, a deterministic optimization is performed in a moving horizon. However, the accuracy of a day-ahead forecast using a one-hour resolution and, therefore, the effectiveness of the optimization is limited. A real-time stochastic programming approach is presented in [17], which can be used to overcome the uncertainty of the PV forecast. Also, in [18], a real-time control is incorporated in an algorithm that tries to maximize the customer satisfaction-involved operational cost while balancing the supply and demand by scheduling EVs, battery storage, grid power, and other flexible loads. Battery degradation is not taken into account here. A range anxiety approach is taken in [11], which penalizes low state of charges (SoC). Here, battery ageing is calculated based on energy throughput. However, no regard has been given to PV/load forecasting. Also, in [12], residential energy costs are minimized with battery degradation taken into account. However, optimized using a one-hour resolution assuming perfect forecasts without adjusting for errors as a result comprising the effectiveness of the optimization. In [19], a method where the charging of EVs at a parking station is controlled based on real-time electricity prices and PV forecast is presented. Here, battery degradation costs are taken into account using a levelized cost of energy approach.

Another important aspect of EV-PV integration in the future smart grid is the provision of ancillary services based on EV storage. This is investigated in [20–27], where fleets of EVs are used as storage and where the scheduling of ancillary services or demand-side management is optimized. However, all of these studies do not take battery/EV degradation into account, and therefore, a significant part of the operating costs is neglected. In addition, scheduling will be skewed after a while when the actual capacity is smaller than taken into account. In [28], the operational costs of V2G are calculated using a simplified battery degradation calculation but are not minimized by the optimization. In [29], an accurate BES degradation model incorporated in a dynamic programming problem is used to optimize the power flows in order to minimize the costs in a PV-EV-BES nano-grid. However, only one BES stress factor is taken into account at the same time. Furthermore, no V2G and no degradation of the EV itself are taken into account. Summarizing the review, it can be concluded that the operational costs of EV/BES are often neglected. PV/load forecasts are only occasionally performed, often in a coarse resolution. Ancillary services are usually only taken into account for larger fleets of EVs, and most papers do not take into account a real-time control scheme. Due to the negligence of costs, coarse resolutions, and lack of error handling mechanisms, the effectiveness of the papers mentioned above is limited.

1.2. Contribution

The main contribution of this study is the combination of several components which previous studies have not combined, together maximizing the effectiveness and accuracy of the proposed solution. This is done by integrating the following:

- **a Lithium-ion degradation model** used to accurately assess and optimize the operational costs of EV and BES. It will be shown that the degradation costs are equal to 68% of the total grid electricity costs and are therefore nonnegligible;
- **a two-stage model predictive controller** consisting of an optimal charging algorithm and real-time controller implemented in a moving horizon control scheme to compensate forecasting and estimation errors, such as PV power or SoC estimation, at a one-minute resolution. It was found that using a moving horizon window and real-time control scheme furthers reduces the costs by 9.7% compared to the reduction in cost of only optimal scheduling;
- **a forecast** of PV power and load demand in 15-min resolution up to 48 h ahead. Even using advanced irradiance forecasting, root mean square errors can be up to 45% [30], showing the necessity of a model predictive controller; and
- **Smart grid implementation** in order for the system to be integrated into a future smart grid, allowing for power curtailment and optimization of available reserved capacity for primary frequency regulation, further reducing the cost by 7.8%.

Using the above components, the actual cost of energy can be accurately assessed and minimized based on forecasts and EV/BES degradation, after which possible errors are compensated and charging can be controlled up to a resolution of one minute while taking into account ancillary services to help increase the EV and PV penetration rate.

1.3. Paper Organization

The paper is organized as follows: In Section 2, the smart charging system and its place in the smart grid paradigm will be elaborated, after which the integration of forecasts will be presented in Section 3. The control scheme will be presented in Section 4. Finally, the obtained results are presented in Section 5.

2. System Description and Smart Grid Implementation

The proposed smart charging system consists of two stages: first, an optimization algorithm finds the optimal charging strategy based on a 15-min resolution. Secondly, a real-time scheduling algorithm operates in real-time within the optimization timesteps. This is integrated into a moving horizon window used to take care of forecast and estimation errors, such as PV/load power, EV arrival times, SoC estimation, etc. The smart charging system is designed to control a multi-port power converter that integrates a PV maximum power point tracker (MPPT), bidirectional BES charger, bidirectional EV charger, and grid-connected inverter on the same DC-link [31,32]; see Figure 1. By connecting these on the same DC-link, several inverting/rectifying power steps can be omitted, achieving higher efficiency and power density. All specifications are given in Table 1; the voltages of the EV and BES are based on [33,34]. Since the inverter is maintaining the power balance on the DC link, it does not require any additional setpoints from the smart charging system. Furthermore, a heat pump connects to the AC side for building heating and tap water. This study assumes that the state-of-charge (SoC) of the BES and EV are known according to the ISO 15118 standard. Finally, as part of the future smart grid, a Smart Grid Operator (SGO) is taken into account, which acts as an aggregator and intermediary between the ancillary services, wholesale market, and small-scale prosumers. This SGO provides the real-time electricity price ($\lambda_{buy/sell}$) as well as up/downregulation prices ($\lambda_{up/dwn}$). As a result, the smart charging algorithm can take place in regulatory services and can take this into account in the optimization. Finally, the SGO can also limit the grid power of the system as part of a demand-side management program, for which the user will be financially compensated afterwards.

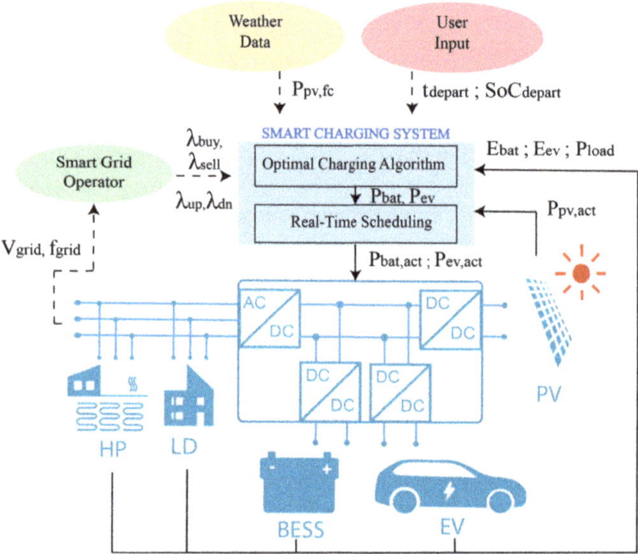

Figure 1. Schematic representation of the system: A multi-port converter including electric vehicle (EV), BES charger, photovoltaïc (PV) maximum power point tracker (MPPT), and grid connected inverter. On the AC side, a heat pump and residential load are connected.

Table 1. Multi-port system parameters.

Symbol	Quantity	Value
P_{PV}^{rated}	installed PV capacity power	10 kWp
P_{ev}	Maximum EV (dis)charging power	10 kW
P	Maximum battery (dis)charging power	10 kW
E_{ev}	Initial full EV capacity	80 kWh
E_{BES}	Initial full battery capacity	10 kWh
$V_{oc,ev}$	EV voltage	325–430 V
$V_{oc,BES}$	BES voltage	325–430 V

3. Second-Life Batteries

Due to the increasing amount of EVs, new markets for second-life EV batteries emerge as EV batteries often have 70–80% remaining capacity left at the end of their EV lifetime [35]. These second-life batteries are then repurposed for stationary applications such as grid reinforcement or demand response systems. This reduces the cost of EV/BES ownership as well as increases the sustainability of the Li-ion batteries. In this study, the second-life value of both the EV and the BES is taken into account and used to assess the operational costs of EV/BES ownership more accurately. Additionally, in [36], it was found that second-life battery performance and state-of-health estimation is strongly influenced by its first-life performance. This motivates the use of a battery degradation model, which minimizes and monitors the degradation such that the performance of the battery in its second-life is increased and more easily assessed.

4. Materials and Methods

The proposed smart charging algorithm can be divided into three subsection: 1. forecasting, 2. optimal scheduling, and 3. moving horizon and real-time control scheme.

4.1. Forecasting

Solar PV and load demand forecasts are required in order to schedule the charging of EV and BES accurately. The forecasting of PV energy can be divided into three categories: statistical data-based methods such as Auto Regressive Integrating Moving Average (ARIMA) [37,38] and machine learning based methods such as neural networks. The disadvantage of the methods mentioned above is that their accuracy solely relies on historical data, as there are no environmental inputs. The third category is a hybrid method based on historical data as well as weather data and satellite images. This study uses the solar radiation forecasts of the royal dutch weather institute. The advantage of this is that the computation of the forecasts is being performed outside the moving horizon controller and is based on previous data, satellite images, weather prediction modelling, and radiative transfer modelling [39]. The forecast is a combination of a short-term forecast (0–6 h) (SEVIRI) and a long-term forecast up to 48 h (HARMONIE) and includes Global Horizontal Irradiance (GHI) and Direct Normal Irradiance (DNI) [30]. In combination with the moving horizon control, good accuracy on the short term can be obtained, while at the same time, scheduling up to 48 h ahead can be performed. Then, using the approach presented in [32], the produced PV power can be calculated based on the orientation of the solar panels. Forecasting the load demand is done based on an aggregated residential profile obtained from [40]. Here, both appliance and heating demand are taken into account. An example of the forecasted and actual powers is shown in Figure 2. Even though the forecasted trend of PV power is accurate, sudden clouding can still cause significant errors of several kilowatts. Similarly, for the load demand profile, its highly stochastic nature is not captured by the aggregated data. However, daily and seasonal variations are incorporated. Due to the existence of these errors combined with a relatively coarse resolution of 15 min in most optimal scheduling problems, significant deviations from the actual optimal solution would occur. This motivates the use of a real-time control scheme.

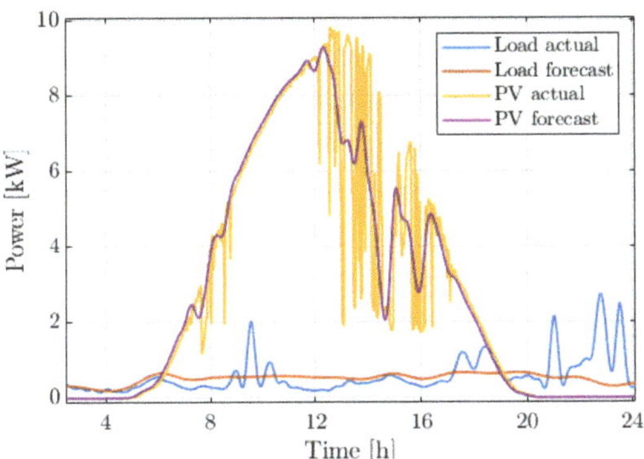

Figure 2. Example of the resulting PV and load forecasts and the actual powers for a summer day.

4.2. Optimal Charging Algorithm

In this section, the optimization model is discussed. All variables are denoted and described in the nomenclature at the end of this article. The optimal charging algorithm's goal is to find the optimal charging schedule based on the forecasts of load and PV while taking into account battery degradation and ancillary services. The main contribution of the proposed optimization problem is the use of a Li-ion battery degradation model. Because of the good balance of power density, energy density, and lifetime, Nickel-Manganese-Cobalt (NMC) based batteries are being used for both vehicle and stationary applications [41]. Therefore, the same ageing model can be used for both the stationary and vehicle battery. Similarly, the costs of V2G, regulatory services, and degradation can be determined based on the actual operating conditions. Furthermore, since the model is included in the objective, the solver will minimize the degradation and the resulting costs. The non-linear behaviour of these cells makes the optimization problem a non-linear programming (NLP) model. The problem was solved using the CONOPT solver (part of BARON/Antigone [42]) using the generic algebraic modelling system (GAMS) platform on a PC with 3.6 GHz, Intel Xeon 4 core and 16 GB RAM.

4.2.1. Objective Function

The objective of this optimization is to minimize the total cost of energy. Here, the total cost of energy C_{total} is made up out of battery energy storage costs C_{BES}, electric vehicle costs C_{EV}, PV energy costs C_{PV}, grid energy costs C_{grid}, and regulatory revenue C_{reg}. Mathematically, this can be expressed as follows:

$$min\left(C_{total}\right) = min\left(C_{BES} + C_{EV} + C_{PV} + C_{grid} - C_{reg}\right) \tag{1}$$

The battery costs are operational costs, which are determined by assessing the remaining value of the degraded battery. This is done by calculating the degraded capacity ΔE_{BES}^{tot} and by subtracting this from the initial capacity E_{BES}^{max}, both in kWh. Next, the remaining value per kWh, V_{BES} in euro/kWh, is calculated according to the model presented in [43]. The decay of V_{BES} versus the remaining capacity is shown in Figure 3. This study assumes that the battery is still in its first life, which ends at 80% remaining capacity, and that the value at the start of second-life V_{BES}^{2nd} equals 50% of a new BES: $V_{BES}^{2nd} = 0.5 V_{BES}^{new}$ [43]. V_{BES} can than be calculated according to the formula presented in Equation (2). The costs are then equal to the difference between a new BES and the degraded BES as shown in Equation (3).

$$V_{BES} = \frac{(V_{BES}^{2nd} - V_{BES}^{new})}{0.2}\Delta E_{BES}^{tot} + V_{BES}^{new} \tag{2}$$

$$C_{BES} = V_{BES}^{new} E_{BES}^{max} - V_{BES}\left(E_{BES}^{max} - \Delta E_{BES}^{tot}\right) \tag{3}$$

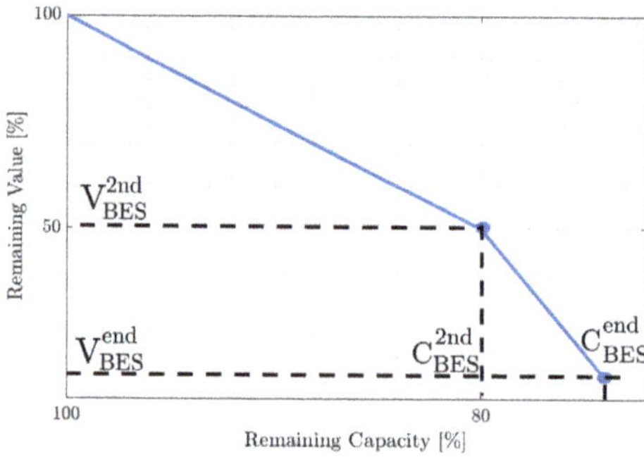

Figure 3. Model of remaining value per kWh as presented in [43]: Here, V_{BES}^{2nd}, V_{BES}^{end}, C_{BES}^{2nd}, and C_{BES}^{end} represent the value at the start of second life, value at the end of its lifetime, capacity at the start of second life, and capacity at the end of lifetime, respectively.

The costs related to the EV are related to the degradation costs of charging and V2G. A similar notation of variables as for the BES is used. Here, the degradation due to driving is not taken into account, as it is not under the control of the smart charging system. Furthermore, it is assumed that the 2nd life of an electric vehicle battery starts at 80% remaining capacity at 50% of its original value per kWh [43]. This is described in Equations (4) and (5).

$$V_{EV} = \frac{(V_{EV}^{2nd} - V_{EV}^{new})}{0.2} \Delta E_{EV}^{tot} + V_{EV}^{new} \tag{4}$$

$$C_{EV} = V_{EV}^{new} E_{EV}^{max} - V_{EV}\left(E_{EV}^{max} - \Delta E_{EV}^{tot} \right) \tag{5}$$

In this study, the cost of PV energy is not assumed to be zero, to take into account the installation and investment costs of the PV system, or to simulate a contractual power purchase agreement. These costs are levelled per kWh to make them independent of the simulation time, without neglecting the related costs. Here, $\lambda_{PV} = 0.03$ €/kWh [12] is assumed. The costs for PV energy are then determined according to the following:

$$C_{PV} = \sum_{t=1}^{T} P_{PV} \Delta t \lambda_{PV} \tag{6}$$

Here, $P_{PV}(t)$ is the generated PV power, Δt is the simulation timestep, and λ_{PV} is the levelized cost of PV energy. The next part of the objective function is the revenue obtained by acting as a frequency containment reserve (FCR) and demand-side management. Here, it is assumed that an SGO acts as an aggregator and mediator between the frequency regulation market and the prosumer, aggregating several flexible systems such that the combined power meets the minimum power requirements for the FCR market. Here, the revenue is obtained by reserving a part of the available power capacity for primary frequency regulation. The up/downregulation prices are $\lambda_{up} and \lambda_{dn}$, respectively. Based on these prices, the operational costs, and the current demand, the optimization will determine how much of the available capacity will be reserved for frequency regulation. The revenue obtained from this is calculated according to the following:

$$C_{reg} = (1 - \epsilon_{fc})\eta_{inv}\eta_{ch} \sum_{t=1}^{T} \left(P_{reg}^{up}(t)\lambda_{up}(t) + P_{reg}^{dwn}(t)\lambda_{dwn}(t) \right) + C_{comp} \tag{7}$$

Here, ϵ_{fc} is the maximum forecasting error; this is taken into account to limit the error between actual and reserved capacity caused by forecasting errors. Due to the moving horizon control, a good short-term accuracy can be achieved. However, as errors will still occur, compensation factor C_{comp} is introduced such that any difference in revenue caused by differences in actual and reserved capacity or in scheduled and used capacity is compensated using C_{comp}. This makes C_{comp} dependent on the bidding process in the regulation market. Therefore, it is assumed that the smart grid operator calculates this compensation factor based on the actual operation. Additionally, C_{comp} could be the compensation obtained for curtailing PV power as part of demand-side management. Here, C_{comp} could be equal to the curtailed energy multiplied by the energy price. Next, $\eta_{inv/ch}$ are the efficiencies of the inverter and BES/EV charger, respectively. Finally, $P_{reg}^{up/dwn}(t)$ are the reserved capacities for regulation. The calculation of these capacities will be explained later in Section 4.2.2.6. The final part of the objective function is the grid energy cost. In this study, a dynamic pricing tariff is used, where the selling price $\lambda_{sell}(t)$ is 10% lower than the buying price $\lambda_{buy}(t)$. This is done to simulate a future environment where electricity prices are a function of demand and supply, optimizing costs and separating buying/selling grid energy results in energy being taken out of the grid when demand and price are low while the energy fed back to the grid is fed in during a time of high demand and high price. The calculation of grid energy costs is shown in Equation (8).

$$C_{grid} = \sum_{t=1}^{T} P_{grid}^{buy}(t)\Delta t \lambda_{buy}(t) - \sum_{t=1}^{T} P_{grid}^{sell}\Delta t \lambda_{sell}(t) \tag{8}$$

Here, $P_{grid}^{buy/sell}(t)$ are the powers drawn and fed from the grid, respectively.

4.2.2. Constraints

4.2.2.1. Lithium-Ion Degradation Model

The battery degradation model is valid for both the EV and the BES, and therefore, the identifier $X = EV = BES$ is used. In order to calculate the capacity lost per time step ($\Delta E_X^{tot}(t)$), the battery degradation model presented in [44] is used. The model is semi-empirical based on 18650 Nickel-Manganese-Cobalt (NMC) cells. It takes into account temperature, current rate ($I_X^{cell}(t)$), and ampere-hours processed ($I_X^{cell}(t)\delta t$). In this study, the cell temperatures are assumed constant at 35 °C, assuming that the EV has a battery temperature control system and because the inside ambient temperature setpoint of the heat pump varies between 18° and 20.5 °C [45]. A distinction between cyclic and calendar ageing is made, denoted as $\Delta E_X^{cycle}(t)$ and $\Delta E_X^{cal}(t)$. Since calendar ageing is mostly dependent on time and temperature [44] and cell temperature is assumed constant, the equation can be simplified to a constant degradation per time step. Here, a minimum lifetime of 5 years is assumed. This is shown in Equation (12).Furthermore, the model is based on the behaviour of a single cell. Therefore the EV/BES power needs to be scaled into the voltage and current of a single cell. To do this, the open-circuit voltage of an NMC cell is used, which can be described by the curve fitted equation shown in Equation (9) [46]. Here, $SoC_x(t)$ denotes the state of charge of x at time t. The resulting curve is shown in Figure 4. Furthermore, it is assumed that $N_X^{parallel}$ by N_X^{series} of these cells are placed in parallel and series respectively, such that the total open-circuit voltage $V_{oc,X}$ resembles existing EV/BES systems [33,34].

$$V_{oc,X}(t) = N_X^{series}\left(a_1 e^{b_1 SoC_x(t)} + a_2 e^{b_2 SoC_x(t)} + a_3 SoC_x(t)^2\right) \qquad ,\forall\, t \tag{9}$$

$$i_X^{cell}(t) = \frac{P_X(t)}{N_X^{parallel} V_{oc,X}(t)} \qquad ,\forall\, t \tag{10}$$

Then, using the calculated cell voltage and current from Equation (9) and (10), the lost capacity per cell can be calculated according to Equations (11) and (13) [44]. Note that the model presented in [44] calculates the percentage of lost charge in ampere hour and is therefore multiplied with $\frac{E_{BES}^{max}}{100}$ in order to get the actually lost capacity. The values of the curve fitted parameters in Equations (9)–(13) can be found in Table 2.

$$\Delta E_X^{cycle}(t) = \left(c_1 e^{c_2 |i_X^{cell}(t)|}|i_X^{cell}(t)|\Delta t\right)\frac{E_{BES}^{max}}{100} \qquad ,\forall\, t \tag{11}$$

$$\Delta E_X^{cal}(t) = \left(c_3 \sqrt{t}\, e^{-24kJ/RT}\right)\frac{E_{BES}^{max}}{100} = \left(c_4\Delta t\right)\frac{E_{BES}^{max}}{100} \qquad ,\forall\, t \tag{12}$$

$$\Delta E_X^{tot} = \sum_{t=0}^{T}\left(\Delta E_X^{cycle}(t) + \Delta E_X^{cal}(t)\right) \tag{13}$$

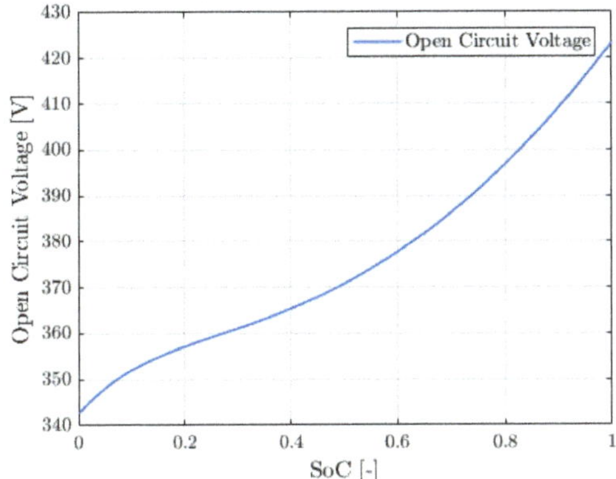

Figure 4. Open circuit voltage for a Nickel–Manganese–Cobalt (NMC) battery cell [46]: Since both the electric vehicle and the stationary battery are made of NMC technology, this cell voltage is applicable for both.

Table 2. Battery model parameters.

Symbol	Quantity	Value
$N_{cell}^{parallel}$	Number of battery cells in parallel	14
N_{cell}^{series}	Number of battery cells in series	100
a_1	$V_{oc}(t)$ curve fit parameter	3.679
b_1	$V_{oc}(t)$ curve fit parameter	−0.1101
a_2	$V_{oc}(t)$ curve fit parameter	−0.2528
b_2	$V_{oc}(t)$ curve fit parameter	−6.829
a_3	$V_{oc}(t)$ curve fit parameter	0.9386
c_1	ageing curve fit parameter	0.00054
c_2	ageing curve fit parameter	0.35
c_3	ageing curve fit parameter	14,876
c_4	averaged calendar ageing per Δt	2.64×10^{-4}

4.2.2.2. Battery Energy Storage Constraints

The BES power $P_{BES}(t)$ is limited between $[-10, 10]$ kW. Additionally, the maximum power $P_{BES}^{max}(t)$ is SoC dependent such that the maximum power drops linearly below a SoC of 10% ($D_{dis} = 0.1$) and above a SoC of 80% ($D_{ch} = 0.8$), representing the constant-current constant-voltage regions of a battery. This is ensured using Equations (14)–(20). The round-trip efficiency is considered constant over power and lifetime and is equal to 95 % [34]. Therefore, the efficiency of a single charge/discharge cycle equals $\eta_{ch/dis} = \sqrt{0.95} = 0.975$. Due to this efficiency, the model recognizes the loss of energy and therefore prevents both $P_{BES}^{neg}(t)$ and $P_{BES}^{pos}(t)$ from having nonzero values at the same time.

$$P_{BES}^{pos}(t) \leq P_{BES}^{max}(t) \qquad , \forall\, t \tag{14}$$

$$P_{BES}^{max}(t) \leq P_{BES}^{rated}(t) \qquad , \forall\, t \tag{15}$$

$$P_{BES}^{max}(t) \leq \frac{P_{BES}^{max}}{1 - D_{ch}} \left(\frac{E_{BES}(t)}{E_{BES}^{max}} - 1 \right) \qquad , \forall\, t \tag{16}$$

$$P_{BES}^{neg}(t) \leq P_{BES}^{min}(t) \qquad , \forall\, t \tag{17}$$

$$P_{BES}^{min}(t) \leq P_{BES}^{rated} \qquad , \forall\, t \tag{18}$$

$$P_{BES}^{min}(t) \leq \frac{P_{BES}^{rated}}{D_{dis}} \frac{E_{BES}(t)}{E_{BES}^{max}} \qquad , \forall\, t \tag{19}$$

$$P_{BES}(t) = \eta_{ch} P_{BES}^{pos}(t) - \frac{1}{\eta_{dis}} P_{BES}^{neg}(t) \qquad , \forall\, t \tag{20}$$

In the above constraints as well as all the following constraints, the superscripts "max", "min", "pos", "neg", and "rated" declare the maximum allowable, minimum allowable, actual positive, actual negative, and rated powers. The energy stored inside the BES $E_{BES}(t)$ can then be calculated according to Equation (21). Here, the BES capacity is fixed for $t = 1$ and $t = t_{final}$ at capacities $E_{BES}^{init} = E_{BES}^{final}$, respectively, in order to have a fair comparison between costs.

$$E_{BES}(t) = \begin{cases} E_{BES}^{init}, & \text{for } t = 1 \\ E_{BES}(t-1) + P_{BES}(t)\Delta t, & \text{for } 1 < t < t_{final} \\ E_{BES}^{final}, & \text{for } t = t_{final} \end{cases} \tag{21}$$

Furthermore, the SoC of the BES is calculated using Equation (22). Here, $E_{BES}^{limit}(t)$ is a variable which represents the actual maximum capacity at time t, which equals the initial maximum capacity E_{BES}^{max} minus the capacity lost by cycling for that time step $\Delta E_{BES}(t)$. This is modelled using Equation (23) and Equation (24).

$$SoC_{BES}(t) = \frac{E_{BES}(t)}{E_{BES}^{limit}(t)} \qquad , \forall\, t \tag{22}$$

$$E_{BES}^{limit}(t) = \begin{cases} E_{BES}^{max}, & \text{for } t = 1 \\ E_{BES}^{limit}(t-1) - \Delta E_{BES}(t), & \text{for } t > 1 \end{cases} \tag{23}$$

$$E_{BES}(t) \leq E_{BES}^{limit}(t) \qquad , \forall\, t \tag{24}$$

4.2.2.3. Electric Vehicle Constraints

The electric vehicle constraints are given in Equations (25)–(36), where the constraints up to Equation (34) are similar to the BES constraints. The electric vehicle is assumed to be unavailable from t_{depart} = 08:00 till t_{arrive} = 18:00 during the day; this is denoted by the binary parameter $EV_{av}(t)$, which equals 0 for $t_{depart} \leq t \leq t_{arrive}$. These departure and arrival times are based on the distribution, as presented in [47]. During that time, it is estimated that the EV is commuting between work and home, where a single trip is 30 km with an efficiency of 15 kWh/100 km [48], resulting in a 9 kWh decrease in charge at arrival; note that the arrival time and charge are only estimations and that possible errors will be compensated after arrival as a result of the moving horizon window.

$$P_{EV}^{pos}(t) \leq P_{EV}^{max}(t) \qquad , \forall\, t \tag{25}$$

$$P_{EV}^{max}(t) \leq P_{EV}^{rated}(t) \qquad , \forall\, t \tag{26}$$

$$P_{EV}^{max}(t) \leq \frac{P_{EV}^{max}}{1 - D_{ch}} \left(\frac{E_{EV}(t)}{E_{EV}^{max}} - 1 \right) \qquad , \forall\, t \tag{27}$$

$$P_{EV}^{neg}(t) \leq P_{EV}^{min}(t) \qquad , \forall\, t \tag{28}$$

$$P_{EV}^{min}(t) \leq P_{EV}^{rated} \qquad , \forall\, t \tag{29}$$

$$P_{EV}^{min}(t) \leq \frac{P_{EV}^{rated}}{D_{dis}} \frac{E_{EV}(t)}{E_{EV}^{max}} \qquad , \forall\, t \tag{30}$$

$$P_{EV}(t) = EV_{av}(t) \left(\eta_{ch} P_{EV}^{pos}(t) - \frac{1}{\eta_{dis}} P_{EV}^{neg}(t) \right) \qquad , \forall\, t \tag{31}$$

$$SoC_{EV}(t) = \frac{E_{EV}(t)}{E_{EV}^{limit}(t)} \qquad , \forall\, t \tag{32}$$

$$E_{EV}^{limit}(t) = \begin{cases} E_{EV}^{max}, \text{for } t = 1 \\ E_{EV}^{limit}(t-1) - \Delta E_{EV}(t), \text{for } t > 1 \end{cases} \tag{33}$$

$$E_{EV}(t) \le E_{EV}^{limit}(t) \qquad , \forall\, t \tag{34}$$

Finally, the user can state a minimum departure charge E_{depart} and departure time t_{depart}, which ensures that the EV always has enough charge at the time of departure. This is modeled using Equation (36):

$$E_{EV}(t) = \begin{cases} E_{EV}^{init}, & \text{for } t = 1 \\ E_{EV}(t-1) + P_{EV}(t)\Delta t, & \text{for } t \le t_{depart} \\ & \&\, t > t_{arrive} \end{cases} \tag{35}$$

$$E_{EV}(t) \ge E_{EV}^{depart}, \text{for } t = t_{depart} \tag{36}$$

4.2.2.4. Power Balance Constraints

For the given system (Figure 1), two power balances exist: 1. on the DC link of the multi-port converter. Here, a positive inverter power $P_{inv}(t)$ equals feeding power to the grid. 2. The second power balance exists on the AC side between the inverter and the meter. This is modeled using Equations (37) and (38).

$$P_{inv}(t) = P_{PV}(t) - P_{BES}(t) - P_{EV}(t) \qquad , \forall\, t \tag{37}$$

$$P_{grid}(t) = \eta_{inv} P_{inv}(t) - P_{load}(t) \qquad , \forall\, t \tag{38}$$

Here, the total load $P_{load}(t)$ consists of the load from all appliances in the building $P_{appl}(t)$ and the power required for heating $P_{heat}(t)$, in the form of a heat pump. In this study, both $P_{appl}(t)$ and $P_{heat}(t)$ are considered non-flexible.

$$P_{load}(t) = P_{appl}(t) + P_{heat}(t) \qquad , \forall\, t \tag{39}$$

4.2.2.5. Grid Constraints

The resulting grid power $P_{grid}(t)$ differentiates between positive and negative grid powers since both have different prices. This is done using the efficiency η_{cable}, which models the power loss in the cable between the meter and converter. This efficiency is assumed to equal 99%. However, more importantly, it ensures that the $P_{grid}^{buy}(t)$ and $P_{grid}^{sell}(t)$ do not have nonzero values simultaneously, as the efficiency loss is recognized. This allows for grid power arbitration without the use of binary variables, drastically increasing the solving time. Here, P_{grid}^{max} is the maximum power of a 3-phase 25A connection.

$$P_{grid}^{sell}(t) \le P_{grid}^{max} \qquad , \forall\, t \tag{40}$$

$$P_{grid}^{buy}(t) \le P_{grid}^{max} \qquad , \forall\, t \tag{41}$$

$$P_{grid}(t) = \eta_{cable} P_{grid}^{sell}(t) - \frac{1}{\eta_{cable}} P_{grid}^{buy}(t) \qquad , \forall\, t \tag{42}$$

4.2.2.6. Regulation Market Constraints

Part of the objective function is the revenue obtained from reserving the capacity for primary frequency regulation. Here, it is assumed that the system is part of a smart grid as described in Section 2. Using Equations (43)–(47), it is ensured that the maximum available capacity for regulation does not exceed the actual maximum capacity in the system. Here, it is important to note that the maximum available EV/BES capacity is SoC dependent. Therefore, the available up/down capacity per converter port $P^x_{up/dwn}(t)$ (x = EV/BES/PV) is denoted using $P^{max/min}_x(t)$, as calculated in Sections 4.2.2.2 and 4.2.2.3. Similarly, it has been done for down regulation in Equations (48)–(53).

$$P^{EV}_{up}(t) \leq EV_{av}(t)\left(P^{min}_{EV}(t) + P_{EV}(t) \right) \qquad , \forall\, t \tag{43}$$

$$P^{BES}_{up}(t) \leq P^{min}_{BES}(t) + P_{BES}(t) \qquad , \forall\, t \tag{44}$$

$$P_{up}(t) \leq \eta_{inv}\left(P^{EV}_{up}(t) + P^{BES}_{up}(t) \right) \qquad , \forall\, t \tag{45}$$

$$P_{up}(t) \leq P^{rated}_{inv}(t) \qquad , \forall\, t \tag{46}$$

$$P_{up}(t) \leq P^{max}_{grid}(t) + P_{load}(t) \qquad , \forall\, t \tag{47}$$

$$P^{EV}_{dwn}(t) \leq EV_{av}(t)\left(P^{max}_{EV}(t) - P_{EV}(t) \right) \qquad , \forall\, t \tag{48}$$

$$P^{BES}_{dwn}(t) \leq P^{min}_{BES}(t) - P_{BES}(t) \qquad , \forall\, t \tag{49}$$

$$P^{PV}_{dwn}(t) \leq P_{PV}(t) \qquad , \forall\, t \tag{50}$$

$$P_{dwn}(t) \leq \eta_{inv}\left(P^{dwn}_{EV}(t) + P^{dwn}_{BES}(t) + P^{dwn}_{PV}(t) \right) \qquad , \forall\, t \tag{51}$$

$$P_{dwn}(t) \leq P^{rated}_{inv}(t) \qquad , \forall\, t \tag{52}$$

$$P_{dwn}(t) \leq P^{max}_{grid}(t) \qquad , \forall\, t \tag{53}$$

In case of symmetric reserve offers, Equation (54) should also be used.

$$P_{up}(t) = P_{dwn}(t) \qquad , \forall\, t \tag{54}$$

4.2.2.7. Inverter Constraints

To account for the efficiency of the inverter, the inverter power is also split into positive and negative parts. The inverter efficiency is assumed equal for both directions of power: 97% and over the entire power range [24].

$$P^{neg}_{inv}(t) \leq P^{max}_{inv} \qquad , \forall\, t \tag{55}$$

$$P^{pos}_{inv}(t) \leq P^{max}_{inv} \qquad , \forall\, t \tag{56}$$

$$P_{inv}(t) = \eta_{inv}P^{pos}_{inv}(t) - \frac{1}{\eta_{inv}}P^{neg}_{inv}(t) \qquad , \forall\, t \tag{57}$$

4.2.2.8 Photovoltaic Constraints

As part of demand-side management as well as for cases with a negative feed-in tariff, it should be possible to curtail PV power. Therefore, in order to allow PV power curtailment, Equation (58) is introduced. Here, the efficiency of the maximum power point tracker is assumed to be $\eta_{mppt} = 98\%$ [49]. Here, $P^{forecast}_{PV}(t)$ denotes the forecasted PV power. This concludes the optimization model section. In the next section, the proposed real-time control is discussed.

$$P_{PV}(t) \leq \eta_{mppt}P^{forecast}_{PV}(t) \qquad , \forall\, t \tag{58}$$

4.3. Moving Horizon Window and Real-Time Control

In Section 4.2, the optimal charging algorithm is discussed. The goal of the optimal charging algorithm is to find the optimal charging schedule within a 24-h optimization window while anticipating future demand and supply based on forecasts and taking into account battery degradation, and primary frequency regulation reserve. However, as shown in Figure 2, the resulting forecasting errors can still be in the range of several kWs. To minimize the effect of these errors, a moving horizon model predictive controller is implemented, which reoptimizes every 15 min (i.e., one optimization timestep). Additionally, a new forecast is obtained every hour to improve accuracy. To compensate for the errors within these 15-min timesteps, such as PV power, SoC estimation, and EV time of arrival, a real-time control scheme is implemented. Here, a rule-based control scheme is implemented such that it can act almost instantly. It should be noted that the real-time control only deals with the errors on top of what is optimally scheduled. For example, if the battery is directly charging from PV power in the optimal solution but the PV power turns out to be less than forecasted, the battery power can be changed accordingly. If no real-time control is implemented, this error needs to be compensated using grid power to maintain the power balance on the DC-link inside the multi-port converter, which could lead to nonoptimal solutions. Figure 5 shows the flowchart of the moving horizon control and the real-time control scheme. First, the irradiance forecasts are obtained, and the resulting PV power is calculated. Then, after the optimization, the output at $t = 1$ is saved and the sample rate is increased 15 times, such that the resolution is now one minute. The interpolated variables are indicated with index k. Furthermore, superscripts fc and act denote the forecasted and actual powers, respectively. In practice, the resolution can be as small as computation time allows to get actual real-time operation. The new real-time charging schedule is determined based on the amplitude of the error, the available EV/BES power, and the current electricity price. The error is calculated according to Equations (59)–(61).

$$\Delta P_{PV}(k) = P_{PV}^{fc}(k) - P_{PV}^{act}(k) \tag{59}$$

$$\Delta P_{load}(k) = P_{load}^{fc}(k) - P_{load}^{act}(k) \tag{60}$$

$$P_{error}(k) = \Delta P_{load}(k) - \Delta P_{PV}(k) \tag{61}$$

Here, a positive error means an excess of power. For example, when the actual *PV* power is higher as forecasted (i.e., $\Delta P_{PV}(k) > 0$) while the actual load is lower as anticipated (i.e., $\Delta P_{load}(k) < 0$), the resulting error is positive $P_{error}(k) > 0$. Next, the power limitations of the EV/BES converter at time k are determined based on power rating, SoC, and power balance inside the multi-port system. This is done using Equations (62)–(65). A rule-based control scheme then determines how the error should be compensated based on P_{error}, the power limitations, current electricity price, and the mean electricity price calculated according to Equation (66), as shown in Figure 5. The goal of the real-time control scheme is to prevent feeding in power to the grid at times of low electricity prices or drawing power during high electricity prices. For example, when the error is positive (meaning an excess of power) and the electricity price λ_{buy} is above average (λ_{mean}), the power is fed to the grid. If the price would be below average, the control first checks whether the BES can absorb the power and, if not, whether the EV can absorb the power; if both are not able to absorb the excess power, it is fed to the grid. After the real-time powers have been calculated, the actual degradation is calculated and the corresponding variables are adjusted. This will then be used for initializing the next optimization instance.

$$P_{BES}^{max,neg}(k) = \max\left(0, \min\left(P_{BES}^{max}, \frac{E_{BES}(k) - E_{BES}^{min}}{\Delta k}, P_{inv}^{max} - P_{PV}(k) + P_{EV}(k)\right)\right) \tag{62}$$

$$P_{BES}^{max,pos}(k) = \max\left(0, \min\left(P_{BES}^{min}, \frac{E_{BES}^{max} - E_{BES}(k)}{\Delta k}, P_{inv}^{max} + P_{PV}(k) - P_{EV}(k)\right)\right) \tag{63}$$

$$P_{EV}^{max,neg}(k) = \max\left(0, \min\left(P_{EV}^{max}, \frac{E_{EV}(k) - E_{EV}^{min}}{\Delta k}, P_{inv}^{max} - P_{PV}(k) + P_{BES}(k)\right)\right) \tag{64}$$

$$P_{EV}^{max,pos}(k) = \max\left(0, \min\left(P_{EV}^{min}, \frac{E_{EV}^{max} - E_{EV}(k)}{\Delta k}, P_{inv}^{max} + P_{PV}(k) - P_{BES}(k)\right)\right) \tag{65}$$

$$\lambda_{mean} = \frac{\sum_{t=t}^{t=T} \lambda_{buy}(t)}{T}, \qquad T \in [t, t + 24h] \tag{66}$$

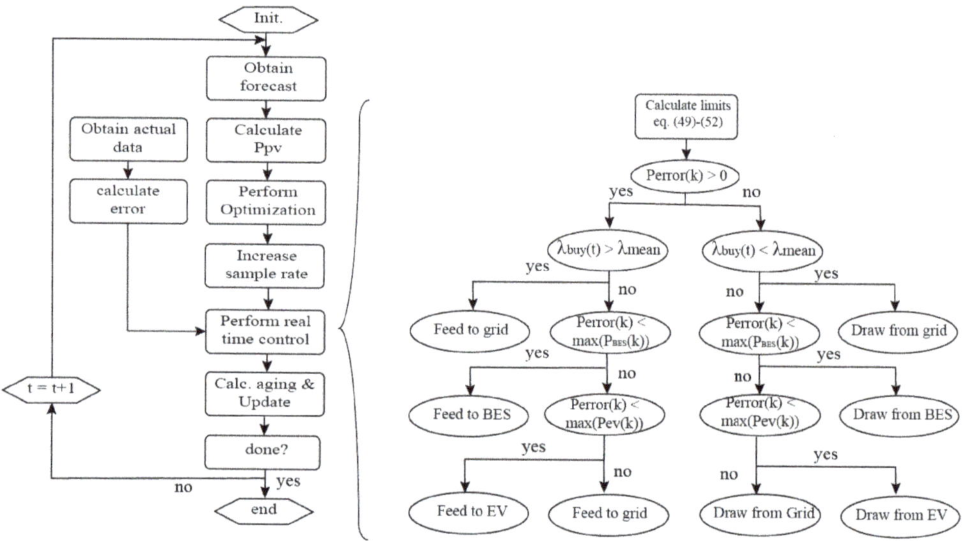

Figure 5. Moving horizon and real-time control flowchart.

5. Use Case and Price Mechanism

The system specifications can be found in Table 1 above. The proposed optimal control scheme can be applied to any building with an EV, solar panels, heat pump, and BES. In this study, a residential building has been chosen as a use case. The building appliance load is obtained from a Dutch distribution system operator and represents a building with a medium to high demand. Both heating and appliance loads are assumed to be non-flexible such that no compromise is given to the user. The power profile of the heat pump is interpolated hourly data obtained from a study performed by the Dutch Organization for Applied Scientific Research [45]. Here, a medium isolated residential building is used with an inside temperature setpoint that varies between 18 °C and 20.5 °C during night and day, respectively. It is assumed that a time-varying electricity price is obtained from the SGO. The price signal is equal to the Amsterdam Power Exchange market [38,50] only averaged around 0.20 €/kWh, see Figure 6. Additionally, up/downregulation prices $\lambda_{up/down}$ are obtained from the SGO and shown in Figure 7. The model has been formulated such that it can operate with any kind of regulation market. In this case, the German primary frequency control market was chosen because it resembles the Dutch frequency control market. The German Frequency Regulation market is a symmetrical market, so both up- and downregulation are equally priced. Figure 7 displays the prices obtained for every week in 2018 [51].

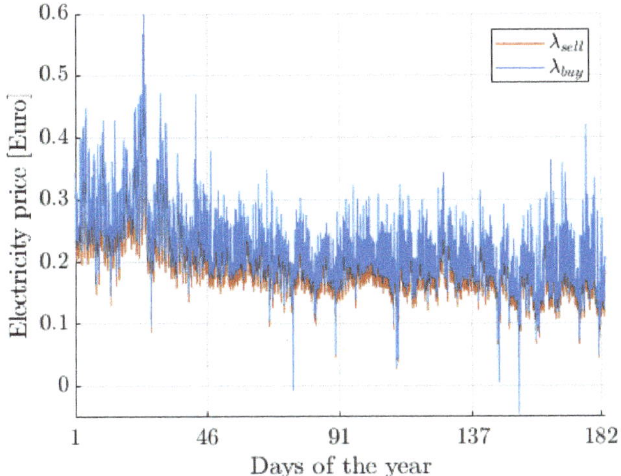

Figure 6. Electricity price based on the 2018 Amsterdam Power Exchange (APX) spot market, averaged around 0.2 €/kWh.

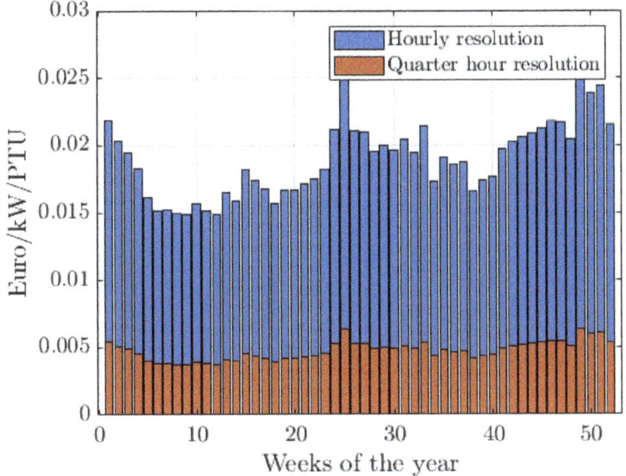

Figure 7. Up- and downregulation prices of the 2018 German market.

6. Results and Discussion

The results for summer and winter days are shown in Figures 8–11, respectively. Here, the electricity buying price and forecasted and actual PV/load powers are included as well. From both Figures 8 and 10, it is clear that the optimal charging stage trades energy in between times of high and low prices. Similarly, it decides when to store PV energy and when to feed it back to the grid. During winter, the energy demand is much higher, mainly because of the increased heating demand. Additionally, the PV production is much lower and, therefore, less energy is fed to the grid. Without an appropriate control scheme, the BES would only be used in case of excess PV energy, therefore resulting in poor utilization of the BES. However, the proposed model still fully utilizes the available BES capacity in order to reduce the cost of energy. This is also shown in Figures 9 and 11. The EV is charged at night above the required departure charge of 50 kWh, such that it can utilize V2G when prices are high in the morning and again in the evening up to the point where prices are low again at the end of the day. Note that the usable EV capacity is not completely utilized, as the inverter power rating limits the possible power exchanged with the grid, whereas outside the instances of peak/valley prices, the difference between feed-in and retail price is probably not enough to overcome the additional losses caused by the degradation of charging at higher powers. The role of the real-time control scheme can be seen in Figure 8; as PV forecasting errors occur, the algorithm decides whether to use the BES, EV, or grid to compensate for these errors. The grid is used for positive errors (excess of power) and high prices, while the BES/EV is used when errors are positive but prices are low (if possible to store energy) and vice versa for high prices and negative errors. Finally, the effect of the degradation model is seen by the peak powers of the EV/BES. Both are rated at 10 kW. However, their powers only exceed the [−6 kW, 6 kW] range at 3.44% of the time (at 1-min resolution).

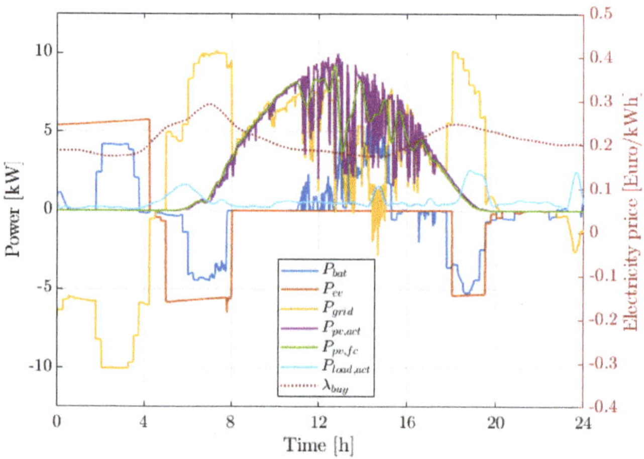

Figure 8. (left axis:) Optimized power flows for a summer day and (right axis:) energy buying price.

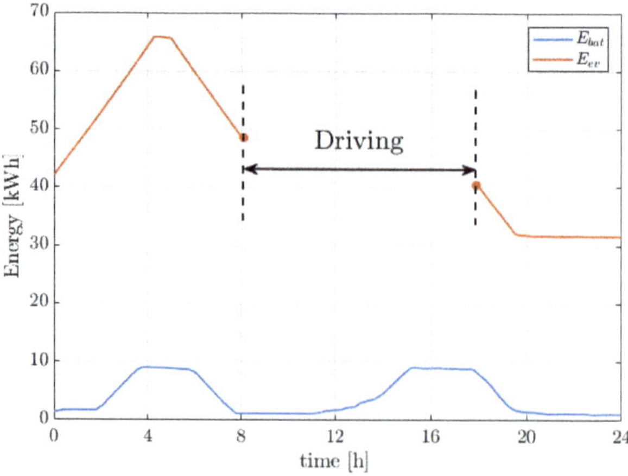

Figure 9. Resulting charge from optimized power flows inside electric vehicle and stationary storage for a summer day.

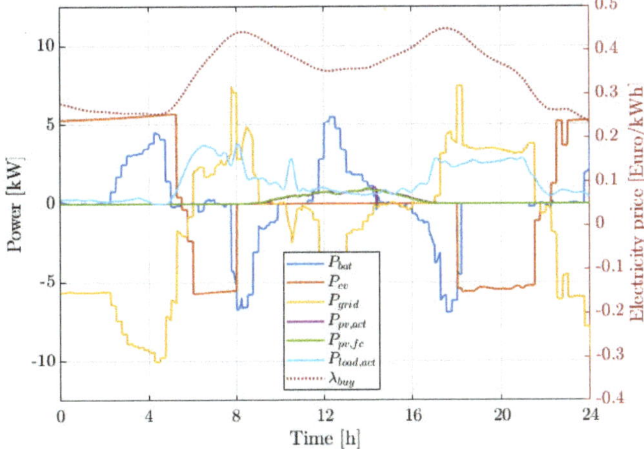

Figure 10. (left axis:) Optimized power flows for a winter day and (right axis:) energy buying price.

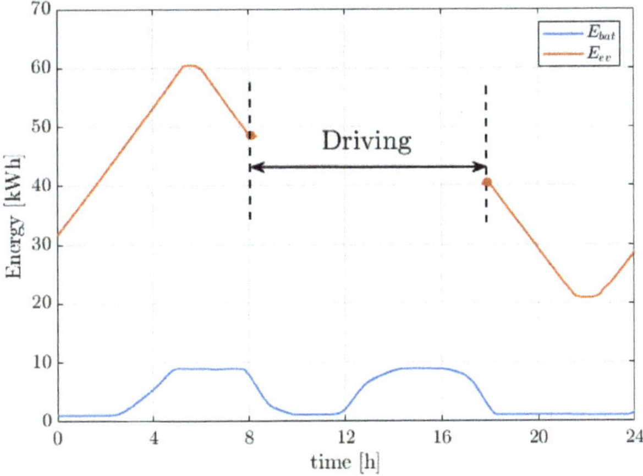

Figure 11. Resulting charge from optimized power flows inside electric vehicle and stationary storage for a winter day.

6.1. Comparison

From the literature review presented above, it can be concluded that most of the existing studies only discuss the optimal scheduling of EV/BES charging. However, within a 15-min forecasting resolution, large forecasting errors can occur due to the fast changing intermittent character of PV power. Additionally, other estimations such as $t_{arrival}$ can lead to more errors. Therefore, a moving horizon window including real-time control scheme is an important part for a smart charging algorithm, as errors are compensated and the optimization is iterated. In order to assess the effectiveness of all the different components, several case studies are performed over a half-year period:

- case 1: Uncontrolled case (only EV, PV, and load)
- case 2: Proposed optimal and real-time control scheme
- case 3: Proposed optimal control and error compensation using grid power
- case 4: Proposed optimal and real-time control scheme without V2G.
- case 5: Proposed optimal and real-time control scheme without up/downregulation.

Here, it is assumed that a half-year simulation period is enough to capture all seasonal variations and that the forecasting error $\epsilon_{fc} = 0$ in order to assess the maximal potential of acting as a FCR. Furthermore, the moving horizon control will operate with a 24-h window in a 15-min resolution. The first uncontrolled case consists only of a PV installation, EV charger, and load. Here, it is assumed that the EV starts charging upon arrival with a 3 kW charger for 3 h to meet its daily demand. The second case is the complete proposed control scheme. The third case is similar to having only an optimal scheduling algorithm; here, all deviations from the obtained optimal solution are compensated using grid energy. The fourth case does not utilize V2G, and the fifth case does not take into account a primary frequency regulation reserve revenue. Figure 12 shows the resulting grid power for use case 2 and 3. Here, case 3 is comparable with having only an optimal control scheme (no real-time control). It can be seen that the errors are dealt with differently, in the end, increasing costs. In Figure 13, the total cumulative costs for every use case over a half year period is shown. Using the proposed control scheme (case 2), the total costs can be reduced by 98.6% compared to the uncontrolled case.

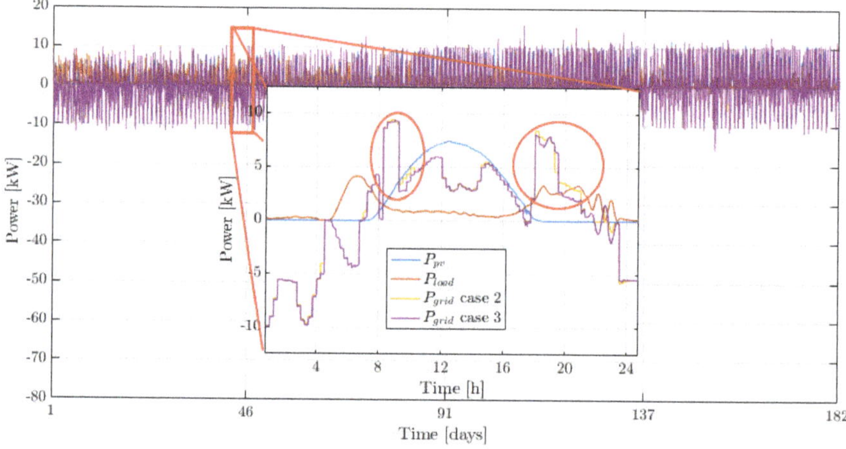

Figure 12. Comparison of grid powers for case 2 and case 3 (with and without real-time control).

A breakdown of all cost components is shown in Figure 14 and Table 3. Here, it can be seen that the optimized EV and BES degradation costs for case 2 are still equal to 91.87 euro and 64.6 euro, respectively. Similarly, the total cost of PV energy equals 168.15 euro. From this, it can be concluded that all these costs are a nonnegligible part of an objective function when minimizing the total cost of energy in an EV-PV-BES-HP system. However, although these costs are relatively high, from Figure 13 and Table 3, it can be concluded that V2G still is a cost-effective method for storing renewable and demand-side management, as the revenue obtained from trading energy exceeds the costs of degradation. Furthermore, it can be seen that the EV charging costs of case 1 are actually the lowest. This is because the average charging power for case 1 is lower compared to the other cases, resulting in lower degradation, however, resulting in more grid electricity costs.

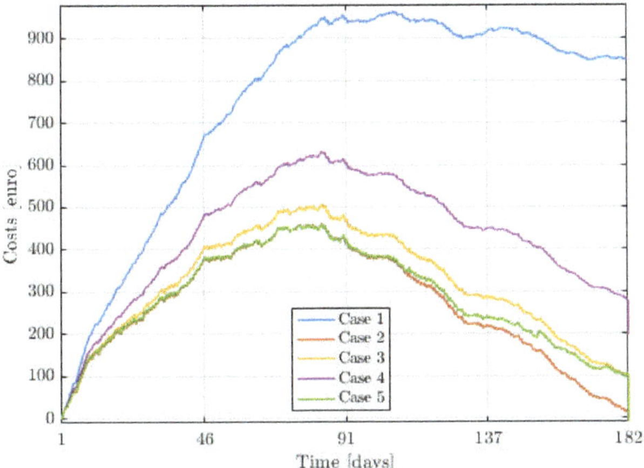

Figure 13. Total costs for all 5 cases over a half-year simulation period.

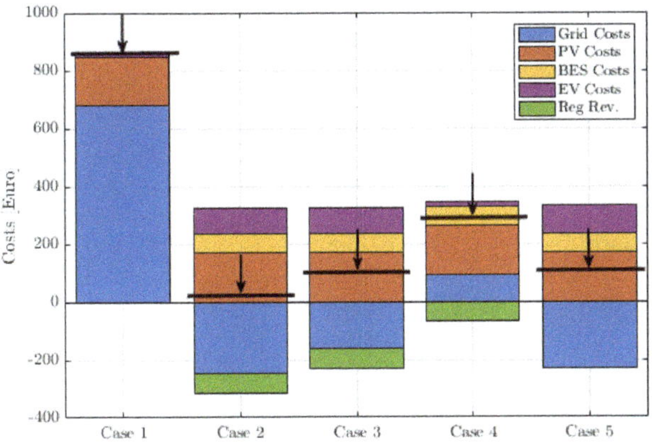

Figure 14. Cost breakdown for all 5 cases: Here, the black line indicates the total costs.

Table 3. Cost comparison of the presented 5 use cases.

Use Case	C_{grid} [€]	C_{PV} [€]	C_{BES} [€]	C_{EV} [€]	C_{Reg} [€]	C_{Total} [€]
1	679.5	169.14	0	10.38	0	859.01
2	-247.27	169.14	64.61	91.88	-66.74	11.76
3	-163.09	169.14	64.34	91.7	-66.74	95.24
4	93.4	169.14	65.68	16.94	-66.83	278.3
5	-230.74	169.14	65.54	98.367	0	102.3

6.1.1. Demand-Side Management: Power Curtailment

Besides operating as a primary frequency regulation reserve, the smart charging algorithm is also capable of power curtailment. For example, when the smart grid operator foresees an over-voltage occurring in the near future, it can choose to limit the grid feed-in power of the system. An example of this is shown in Figure 15. Here, the SGO reduces the maximum allowable feed-in power to 5 kW between 09:30 and 18:00. A comparison with the same day without power curtailment is shown in Figure 16.

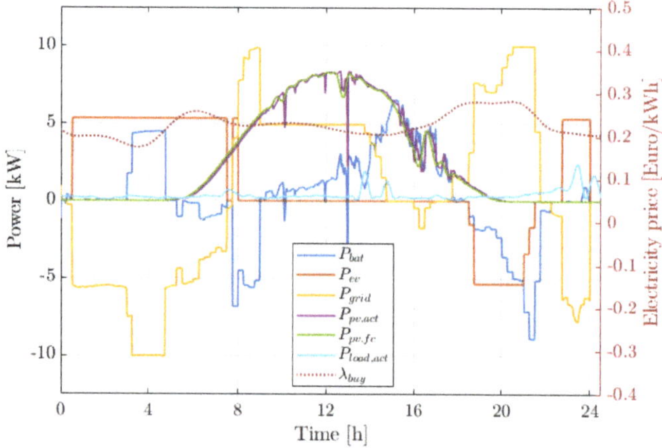

Figure 15. Curtailment of PV power due to reduced maximum allowable grid feed-in between 08:00 and 18:00.

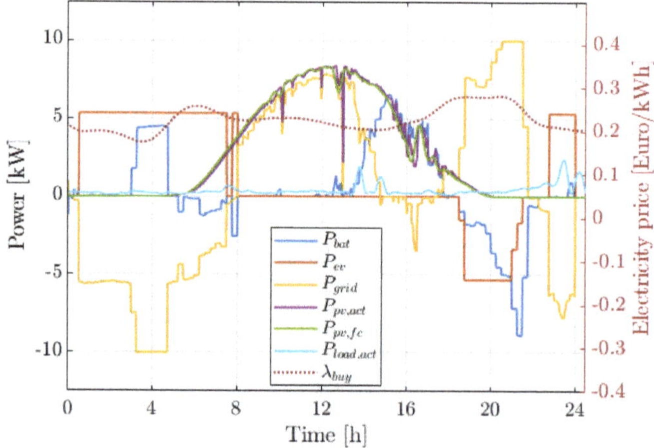

Figure 16. Comparison of the same day without power curtailment.

7. Conclusions

In this paper, a building smart charging algorithm was presented for a multi-port system integrating EV, PV, BES, and a HP. Here, the HP and appliance load were assumed fixed. Then, based on forecasts of PV production and load, the smart charging algorithm minimizes the total cost of energy incorporating grid electricity costs, PV investment and installation costs, EV/BES operational costs, and revenue obtained from operating as primary frequency regulation reserve. It has been shown that the proposed algorithm is very effective as it reduced 98.6% of the total cost of energy compared to an uncontrolled EV-PV-HP case. Additionally, the potential of V2G and the importance of forecasting error handling using a real-time moving horizon control scheme are discussed. Finally, it has been shown that EV/BES degradation costs as well as PV investment/installation costs are nonnegligible parts of an objective function which tries to minimize the total cost of energy in an EV-PV-BES-HP system.

Author Contributions: Conceptualization, W.V., G.R.C.M., and P.B.; formal analysis, W.V. and G.R.C.M.; funding acquisition, G.R.C.M. and P.B.; investigation, W.V.; methodology, W.V. and G.R.C.M.; project administration, G.R.C.M. and P.B.; resources, W.V., G.R.C.M., and P.B.; software, W.V.; supervision, G.R.C.M. and P.B.; validation, W.V. and G.R.C.M.; visualization, W.V.; writing—original draft, W.V.; writing—review and editing, W.V. and G.R.C.M. All authors have read and agreed to the published version of the manuscript.

Funding: This research was funded by the FLEXgrid project, part of TKI urban energy, with contributions from Stedin, Alfen, and Power Research Electronics.

Acknowledgments: The authors would like to acknowledge the project partners at Power Research Electronics, Stedin and Alfen, and the GAMS support team for their prompt responses and help in the project.

Conflicts of Interest: The sponsors had no role in the design of the study; in the collection, analyses, or interpretation of data; in the writing of the manuscript; or in the decision to publish the results.

Abbreviations

The following abbreviations are used in this manuscript:

t	optimization time index (-)
Δt	optimization time step (-)
k	real-time control time index (-)
C_{total}	Total cost of energy (Euro)
C_{BES}	BES costs (Euro)
C_{EV}	Electric vehicle costs (Euro)
C_{PV}	PV costs (Euro)
C_{grid}	Grid energy costs (Euro)
C_{reg}	up/downregulation revenue (Euro)
V_{BES}^{new}	New BES price per kWh (500 Euro)
V_{new}^{EV}	New EV price per kWh (500 Euro)
V_{BES}^{2nd}	2nd life BES price per kWH (250 Euro)
V_{EV}^{2nd}	2nd life EV price per kWH (250 Euro)
λ_{PV}	PV energy price (0.03 Euro)
λ_{buy}	Grid energy buying price (Euro)
λ_{sell}	Grid energy buying price (Euro)
λ_{up}	up regulation price (Euro)
λ_{down}	down regulation price (Euro)
$E_{BES}(t)$	BES capacity at time t (kWh)
E_{BES}^{max}	Initial maximum BES capacity (10 kWh)
$E_{BES}^{limit}(t)$	Max BES capacity at time t (kWh)
E_{BES}^{init}	Initial BES capacity (5 kWh)
ΔE_{BES}^{tot}	Total degraded BES capacity (kWh)
ΔE_{EV}^{tot}	Total degraded EV capacity (kWh)
$E_{EV}(t)$	EV capacity at time t (kWh)
E_{EV}^{max}	Initial maximum EV capacity (80 kWh)
$E_{EV}^{limit}(t)$	Max EV capacity at time t (kWh)
E_{EV}^{init}	Initial EV capacity (35 kWh)
E_{EV}^{final}	Final EV capacity (35 kWh)
E_{EV}^{depart}	EV departure charge (50 kWh)
$P_{inv}(t)$	inverter power at time t (kW)
P_{inv}^{max}	Max inverter power (10 kW)
P_{inv}^{neg}	negative (draw) inverter power (kW)
P_{inv}^{pos}	positive (feed-in) inverter power (kW)
$P_{BES}(t)$	BES power at time t (kW)
P_{BES}^{max}	Max BES power (10 kW)
P_{BES}^{neg}	discharging BES power (kW)
P_{BES}^{pos}	charging BES power (kW)
$P_{EV}(t)$	EV power at time t (kW)
P_{EV}^{max}	Max EV power (kW)
P_{EV}^{neg}	discharging EV power (kW)
P_{EV}^{pos}	charging EV power (kW)
$P_{PV}(t)$	produced PV power at time t (kW)
$P_{PV}^{forecast}(t)$	forecasted PV power at time t (kW)

$P_{PV}^{act}(t)$	forecasted PV power at time t (kW)
$P_{load}(t)$	total load power at time t (kW)
P_{grid}	grid power at time t (kW)
P_{grid}^{max}	Maximum grid power (kW)
P_{grid}^{sell}	feed-in grid power (kW)
P_{grid}^{buy}	buying grid power (kW)
P_{reg}^{up}	available up-regulation capacity (kW)
P_{reg}^{dwn}	available down-regulation capacity (kW)
V_{BES}^{cell}	Average BES cell voltage (3.7 V)
$V_{oc,BES}$	open circuit voltage BES (V)
$V_{oc,EV}$	open circuit voltage BES (V)
I_{BES}^{cell}	BES cell current (A)
I_{EV}^{cell}	BES cell current (A)
SoC_{BES}	BES Energy Storage State of Charge (-)
SoC_{EV}	Electric vehicle State of Charge (-)
η_{mppt}	MPPT efficiency (98%)
η_{inv}	Inverter efficiency (96%)
η_{ch}	EV/BES charging efficiency (97.5%)
η_{dis}	EV/BES discharging efficiency (97.5%)
η_{cable}	cable efficiency (99%)
t_{depart}	Electric vehicle departure time (8:00 h)
$interest$	Bank account interest rate (1%/year)
$a1$	$V_{oc}(t)$ a1 (3.679)
$a2$	$V_{oc}(t)$ a2 (-0.2528)
$a3$	$V_{oc}(t)$ a3 (0.9386)
$b1$	$V_{oc}(t)$ b1 (-0.1101)
$b2$	$V_{oc}(t)$ b2 (-6.829)
$c1$	$V_{oc}(t)$ c (0.00054)
$c2$	$V_{oc}(t)$ c (0.35)
$c3$	$V_{oc}(t)$ c (14,876)
$c3$	$V_{oc}(t)$ c (2.64×10^{-4})
$N_{EV}^{parallel}$	Amount of EV battery cell groups in parallel (145)
N_{EV}^{series}	Amount of EV battery cell in series (100)
N_{BES}^{series}	Amount of cells in series in BES (100)
$N_{BES}^{parallel}$	Amount of cells in parallel in BES [18]

References

1. Global Transportation Energy Consumption: Examination of Scenarios to 2040 Using ITEDD. Available online: https://www.eia.gov/analysis/studies/transportation/scenarios/pdf/globaltransportation.pdf (accessed on 7 Apri 2020).
2. Kempton, W.; Tomić, J. Vehicle-to-grid power implementation: From stabilizing the grid to supporting large-scale renewable energy. *J. Power Sources* **2005**, *144*, 280–294. [CrossRef]
3. Birnie, D.P. Solar-to-vehicle (S2V) systems for powering commuters of the future. *J. Power Sources* **2009**, *186*, 539–542. [CrossRef]
4. haikh, P.H.; Nor, N.B.M.; Nallagownden, P.; Elamvazuthi, I.; Ibrahim, T. A review on optimized control systems for building energy and comfort management of smart sustainable buildings. *Renew. Sustain. Energy Rev.* **2014**, *34*, 409–429. [CrossRef]
5. Cao, X.; Dai, X.; Liu, J. Building Energy-Consumption Status Worldwide and the State-of-the-Art Technologies for Zero-Energy Buildings during the Past Decade. *Energy Build.* **2016**, *128*, 198–213. [CrossRef]
6. Teleke, S.; Baran, M.E.; Bhattacharya, S.; Huang, A.Q. Rule-Based Control of Battery Energy Storage for Dispatching Intermittent Renewable Sources. *IEEE Trans. Sustain. Energy* **2010**, *1*, 117–124. [CrossRef]
7. Doukas, H.; Patlitzianas, K.D.; Iatropoulos, K.; Psarras, J. Intelligent building energy management system using rule sets. *Build. Environ.* **2007**, *42*, 3562–3569. [CrossRef]

8. Wi, Y.-M.; Lee, J.-U.; Joo, S.-K. Electric vehicle charging method for smart homes/buildings with a photovoltaic system. *IEEE Trans. Consum. Electron.* **2013**, *59*, 323–328. [CrossRef]

9. Brahman, F.; Honarmand, M.; Jadid, S. Optimal electrical and thermal energy management of a residential energy hub, integrating demand response and energy storage system. *Energy Build.* **2015**, *90*, 65–75. [CrossRef]

10. Bozchalui, M.C.; Hashmi, S.A.; Hassen, H.; Canizares, C.A.; Bhattacharya, K. Optimal Operation of Residential Energy Hubs in Smart Grids. *IEEE Trans. Smart Grid* **2012**, *3*, 1755–1766. [CrossRef]

11. Igualada, L.; Corchero, C.; Cruz-Zambrano, M.; Heredia, F.-J. Optimal Energy Management for a Residential Microgrid Including a Vehicle-to-Grid System. *IEEE Trans. Smart Grid* **2014**, *5*, 2163–2172. [CrossRef]

12. Sun, Y.; Yue, H.; Zhang, J.; Booth, C. Minimization of Residential Energy Cost Considering Energy Storage System and EV with Driving Usage Probabilities. *IEEE Trans. Sustain. Energy* **2019**, *10*, 1752–1763. [CrossRef]

13. Pedrasa, M.A.A.; Spooner, T.D.; MacGill, I.F. Coordinated Scheduling of Residential Distributed Energy Resources to Optimize Smart Home Energy Services. *IEEE Trans. Smart Grid* **2010**, *1*, 134–143. [CrossRef]

14. Tushar, W.; Yuen, C.; Huang, S.; Smith, D.B.; Poor, H.V. Cost Minimization of Charging Stations with Photovoltaics: An Approach with EV Classification. *IEEE Trans. Intell. Transp. Syst.* **2016**, *17*, 156–169. [CrossRef]

15. Zhang, J.; Zhang, Y.; Li, T.; Jiang, L.; Li, K.; Yin, H.; Ma, C. A Hierarchical Distributed Energy Management for Multiple PV-Based EV Charging Stations. In Proceedings of the IECON 2018—44th Annual Conference of the IEEE Industrial Electronics Society, Washington, DC, USA, 21–23 October 2018; IEEE: Washington, DC, USA, 2018; pp. 1603–1608.

16. Guo, Y.; Xiong, J.; Xu, S.; Su, W. Two-Stage Economic Operation of Microgrid-Like Electric Vehicle Parking Deck. *IEEE Trans. Smart Grid* **2016**, *7*, 1703–1712. [CrossRef]

17. Liu, Z.; Wu, Q.; Shahidehpour, M.; Li, C.; Huang, S.; Wei, W. Transactive Real-Time Electric Vehicle Charging Management for Commercial Buildings with PV On-Site Generation. *IEEE Trans. Smart Grid* **2019**, *10*, 4939–4950. [CrossRef]

18. Yan, Q.; Zhang, B.; Kezunovic, M. Optimized Operational Cost Reduction for an EV Charging Station Integrated with Battery Energy Storage and PV Generation. *IEEE Trans. Smart Grid* **2019**, *10*, 2096–2106. [CrossRef]

19. Chaudhari, K.; Ukil, A.; Kumar, K.N.; Manandhar, U.; Kollimalla, S.K. Hybrid Optimization for Economic Deployment of ESS in PV-Integrated EV Charging Stations. *IEEE Trans. Ind. Inf.* **2018**, *14*, 106–116. [CrossRef]

20. Tran, V.T.; Islam, M.d.R.; Muttaqi, K.M.; Sutanto, D. An Efficient Energy Management Approach for a Solar-Powered EV Battery Charging Facility to Support Distribution Grids. *IEEE Trans. Ind. Appl.* **2019**, *55*, 6517–6526. [CrossRef]

21. Weckx, S.; Driesen, J. Load Balancing with EV Chargers and PV Inverters in Unbalanced Distribution Grids. *IEEE Trans. Sustain. Energy* **2015**, *6*, 635–643. [CrossRef]

22. O'Connell, A.; Flynn, D.; Keane, A. Rolling Multi-Period Optimization to Control Electric Vehicle Charging in Distribution Networks. *IEEE Trans. Power Syst.* **2014**, *29*, 340–348. [CrossRef]

23. Singh, M.; Kar, I.; Kumar, P. Influence of EV on grid power quality and optimizing the charging schedule to mitigate voltage imbalance and reduce power loss. In Proceedings of the 14th International Power Electronics and Motion Control Conference EPE-PEMC 2010, Ohrid, Macedonia, 6–8 September 2010; pp. T2-196–T2-203.

24. Chandra Mouli, G.R. Charging Electric Vehicles from Solar Energy: Power Converter, Charging Algorithm and System Design. Ph.D. Thesis, Delft University of Technology, Delft, The Netherlands, 2018.

25. Hu, J.; You, S.; Lind, M.; Ostergaard, J. Coordinated Charging of Electric Vehicles for Congestion Prevention in the Distribution Grid. *IEEE Trans. Smart Grid* **2014**, *5*, 703–711. [CrossRef]

26. Mets, K.; Verschueren T.; De Turck F.; Develder C. Exploiting V2G to optimize residential energy consumption with electrical vehicle (dis)charging. In Proceedings of the IEEE First International Workshop on Smart Grid Modeling and Simulation (SGMS 2011), Brussels, Belgium, 17 October 2011.

27. Chen, Q.; Wang, F.; Hodge, B.-M.; Zhang, J.; Li, Z.; Shafie-Khah, M.; Catalao, J.P.S. Dynamic Price Vector Formation Model-Based Automatic Demand Response Strategy for PV-Assisted EV Charging Stations. *IEEE Trans. Smart Grid* **2017**, *8*, 2903–2915. [CrossRef]

28. Sortomme, E.; El-Sharkawi, M.A. Optimal Scheduling of Vehicle-to-Grid Energy and Ancillary Services. *IEEE Trans. Smart Grid* **2012**, *3*, 351–359. [CrossRef]

29. Badawy, M.O.; Sozer, Y. Power Flow Management of a Grid Tied PV-Battery System for Electric Vehicles Charging. *IEEE Trans. Ind. Applicat.* **2017**, *53*, 1347–1357. [CrossRef]

30. Wang, P.; van Westrhenen, R.; Meirink, J.F.; van der Veen, S.; Knap, W. Surface solar radiation forecasts by advecting cloud physical properties derived from Meteosat Second Generation observations. *Solar Energy* **2019**, *177*, 47–58. [CrossRef]

31. Cecati, C.; Khalid, H.A.; Tinari, M.; Adinolfi, G.; Graditi, G. DC nanogrid for renewable sources with modular DC/DC LLC converter building block. *IET Power Electron.* **2017**, *10*, 536–544. [CrossRef]

32. Chandra Mouli, G.R.; Bauer, P.; Zeman, M. System design for a solar powered electric vehicle charging station for workplaces. *Appl. Energy* **2016**, *168*, 434–443. [CrossRef]

33. Tesla Model S. Available online: http://www.roperld.com/science/teslamodels.htm (accessed on 15 May 2020).

34. LG Chem RESU 10H—400V Lithium-Ion Storage Battery. Available online: https://www.europe-solarstore.com/lg-chem-resu-10h-400v-lithium-ion-storage-battery.html (accessed on 15 May 2020).

35. Cready, E.; Lippert, J.; Pihl, J.; Weinstock, I.; Symons, P. *Technical and Economic Feasibility of Applying Used EV Batteries in Stationary Applications*; SAND2002-4084; Sandia National Labs.: Albuquerque, NM, USA, 2003; p. 809607.

36. Martinez-Laserna, E.; Sarasketa-Zabala, E.; Villarreal Sarria, I.; Stroe, D.-I.; Swierczynski, M.; Warnecke, A.; Timmermans, J.-M.; Goutam, S.; Omar, N.; Rodriguez, P. Technical Viability of Battery Second-life: A Study From the Ageing Perspective. *IEEE Trans. Ind. Applicat.* **2018**, *54*, 2703–2713. [CrossRef]

37. Akhter, M.N.; Mekhilef, S.; Mokhlis, H.; Mohamed Shah, N. Review on forecasting of photovoltaic power generation based on machine learning and metaheuristic techniques. *IET Renew. Power Gener.* **2019**, *13*, 1009–1023. [CrossRef]

38. van der Meer, D.; Ram Chandra Mouli, G.; Morales-Espana, G.; Ramirez Elizondo, L.; Bauer, P. Erratum to Energy Management System with PV Power Forecast to Optimally Charge EVs at the Workplace. *IEEE Trans. Ind.* **2018**, *14*, 311–320, 3298. [CrossRef]

39. Forecast of Solar Radiation in The Netherlands. Available online: https://www.knmi.nl/research/observations-data-technology/projects (accessed on 7 April 2020).

40. Voulis, N. Harnessing Heterogeneity: Understanding Urban Demand to Support the Energy Transition. Ph.D. Thesis, Delft University of Technology, Delft, The Netherlands, 2019.

41. Laslau, C.; Xie, L.; Robinson, C. The Next-Generation Battery Roadmap: Quantifying How Solid-State, Lithium-Sulfur, and Other Batteries Will Emerge After 2020; Lux Research Energy Storage Intelligence Research. 2015. Available online: https://members.luxresearchinc.com/research/report/17977 (accessed on 7 April 2020) .

42. BARON. Available online: https://minlp.com/baron (accessed on 7 April 2020).

43. Battery Value Versus Remaining Energy. Available online: https://www.nrel.gov/docs/fy16osti/66140.pdf (accessed on 7 April 2020).

44. Wang, J.; Purewal, J.; Liu, P.; Hicks-Garner, J.; Soukazian, S.; Sherman, E.; Sorenson, A.; Vu, L.; Tataria, H.; Verbrugge, M.W. Degradation of lithium ion batteries employing graphite negatives and nickel–cobalt–manganese oxide + spinel manganese oxide positives: Part 1, aging mechanisms and life estimation. *J. Power Sources* **2014**, *269*, 937–948. [CrossRef]

45. De Systeemkosten van Warmte Voor Woningen. Availabe online: https://repository.tudelft.nl/view/tno/uuid:9d49b1d3-107b-4f49-a36c-b606786bc207 (accessed on 7 April 2020).

46. Baccouche, I.; Jemmali, S.; Manai, B.; Omar, N.; Amara, N. Improved OCV Model of a Li-Ion NMC Battery for Online SOC Estimation Using the Extended Kalman Filter. *Energies* **2017**, *10*, 764. [CrossRef]

47. Analyses. Available online: https://platform.elaad.io/analyses.html (accessed on 7 April 2020).

48. Tesla Model 3 Standard Range. Available online: https://ev-database.org/car/1060/Tesla-Model-3-Standard-Range (accessed on 7 April 2020).

49. Chandra Mouli, G.R.; Schijffelen, J.H.; Bauer, P.; Zeman, M. Design and Comparison of a 10-kW Interleaved Boost Converter for PV Application Using Si and SiC Devices. *IEEE J. Emerg. Sel. Topics Power Electron.* **2017**, *5*, 610–623. [CrossRef]

50.   Market Data. Available online: https://www.epexspot.com/en/market-data (accessed on 7 April 2020).
51.   Regelleistung.Net-Datencenter. Available online: https://www.regelleistung.net/apps/datacenter/tenders/ (accessed on 7 April 2020).

Article

Optimized Scheduling of EV Charging in Solar Parking Lots for Local Peak Reduction under EV Demand Uncertainty

Rishabh Ghotge [1], Yitzhak Snow [1], Samira Farahani [2], Zofia Lukszo [2] and Ad van Wijk [1,*]

[1] Department of Process and Energy, Faculty of Mechanical, Maritime and Materials Engineering, Delft University of Technology, Mekelweg 5, 2628 CD Delft, The Netherlands; r.ghotge@tudelft.nl (R.G.); yitzijansnow@gmail.com (Y.S.)

[2] Department of Engineering Systems and Services, Faculty of Technology, Policy and Management, Delft University of Technology, Jaffalaan 5, 2628 BX Delft, The Netherlands; samifarahani@gmail.com (S.F.); Z.Lukszo@tudelft.nl (Z.L.)

* Correspondence: a.j.m.vanwijk@tudelft.nl; Tel.: +31-627021501

Received: 20 February 2020; Accepted: 6 March 2020; Published: 10 March 2020

Abstract: Scheduled charging offers the potential for electric vehicles (EVs) to use renewable energy more efficiently, lowering costs and improving the stability of the electricity grid. Many studies related to EV charge scheduling found in the literature assume perfect or highly accurate knowledge of energy demand for EVs expected to arrive after the scheduling is performed. However, in practice, there is always a degree of uncertainty related to future EV charging demands. In this work, a Model Predictive Control (MPC) based smart charging strategy is developed, which takes this uncertainty into account, both in terms of the timing of the EV arrival as well as the magnitude of energy demand. The objective of the strategy is to reduce the peak electricity demand at an EV parking lot with PVarrays. The developed strategy is compared with both conventional EV charging as well as smart charging with an assumption of perfect knowledge of uncertain future events. The comparison reveals that the inclusion of a 24 h forecast of EV demand has a considerable effect on the improvement of the performance of the system. Further, strategies that are able to robustly consider uncertainty across many possible forecasts can reduce the peak electricity demand by as much as 39% at an office parking space. The reduction of peak electricity demand can lead to increased flexibility for system design, planning for EV charging facilities, deferral or avoidance of the upgrade of grid capacity as well as its better utilization.

Keywords: electric vehicle; demand forecasting; peak shaving; smart charging; robust optimization

1. Introduction

Globally, road transportation accounts for 17% of all emissions of carbon dioxide (CO_2) [1]. Electric vehicles (EVs) offer a solution for the reduction of emissions in the road transport sector, particularly for passenger vehicles. Two characteristics of EVs already make a convincing case for their adoption: (1) the high efficiencies of electric propulsion and (2) lower or zero tailpipe emissions.

The net CO_2 emissions per kilometer driven by Battery Electric Vehicles (BEVs), however, depend on the energy mix used for electricity generation. Based on Well-to-Wheel comparison, the use of BEVs can greatly reduce transport-related net emissions when they are powered by electricity generated from renewable sources [2]. There is thus a need both to shift road transportation toward electric propulsion as well as to simultaneously increase the renewable fraction of the electricity used to power it.

Currently, the charging of the majority of EVs is uncoordinated or unscheduled, i.e., they begin charging at the moment when they are plugged in. Unscheduled charging of electric vehicles can cause increased demand for electricity at peak times. These peaks are expected to increase to over 50% even at 30% EV penetration in the Netherlands [3]. Higher peak loads lead to local issues such as overloading of transformers and other infrastructure in distribution networks, increased grid congestion, power imbalances and voltage dips [4,5]. At a more global level, higher peak loads can result in higher electricity costs and greater carbon emissions [6].

Scheduled or smart charging of EVs can greatly reduce the peak demand for electricity and avoid local congestion in electrical power systems [7]. This reduces the costs for the provision of ubiquitous and affordable EV charging facilities. In this manner, it lowers one of the main barriers to EV adoption: a lack of accessible charge points. In addition, by matching EV charging with the availability of locally produced renewable energy (such as that produced by solar photovoltaics), smart charging can also result in increasing the penetration of renewables in the mobility sector. EV charge scheduling strategies, which aim to reduce local peak demands and congestion, require knowledge of future electricity demands and renewable energy production. However, this uncertainty remains either neglected or seldom addressed in the literature on smart charging.

In [8], the impact of EV charging on residential distribution grids is investigated, but although the household demands are forecasted stochastically, the arrival of other EVs in the future are not considered. The power demand of a single EV is considerably larger than the loads in the household profiles considered in the study and the energy required for a daily charging session is in the range of a household daily energy demand [9]. The lack of EV load forecasting is thus a considerable oversight—particularly from a peak shaving perspective. In [10], a fleet of V2G compatible EVs is considered, whose scheduling is to be optimized for the purpose of providing spinning reserves. Although the formulation acknowledges and accounts for the unexpected departure of EVs, it assumes that the aggregator has accurate information on both the EV driving patterns as well as their States of Charge (SoCs), based on which demands are calculated.

In [11], EVs are scheduled for peak reduction and self-consumption within a microgrid. A car-sharing setup is considered, where the users reserve vehicles in advance for the trips they plan. In such a case, the deviation from the planned schedule is small, assumed to be always less than an hour. EVs that are not used in such a car-sharing scheme are not considered. Similarly, in [12], fuel cell electric vehicles are scheduled for V2G energy dispatch in a microgrid. However, although load forecasting is performed with an assumption of accuracy, mobility-related uncertainty associated with the future arrival of vehicles remains unaddressed.

Neglect or incomplete consideration of future EV demand in these models can cause smart charging strategies to perform worse than expected. When designing an optimal strategy, it is critical that uncertainty of vehicle charging demand (both in terms of timing as well as magnitude) is taken into account. This work investigates and quantifies the effect of this uncertainty. It is taken into consideration to develop strategies based on Model Predictive Control (MPC) for scheduling EV charging.

The paper is divided into sections as follows: Section 2 describes the physical system considered i.e., the solar parking lot, EVs and Electric Vehicle Supply Equipment (EVSE) and its modeling. Section 3 introduces the proposed methods of charge scheduling based on MPC methodology and their formulation. Section 4 illustrates the results obtained from running the simulations and discusses their relevance. Finally, Section 5 provides the conclusions and insights provided by the paper together with interesting directions for future research.

2. System Description

The system considered in this work is a solar charging carport for the charging of electric vehicles as seen in Figure 1. It was modeled in MATLAB and was used to generate inputs for the scheduling strategy. It includes a solar photovoltaic (PV) array, stationary storage and LED lighting connected to a DC bus, coupled bidirectionally with a grid-connected AC bus, which also enables AC charging.

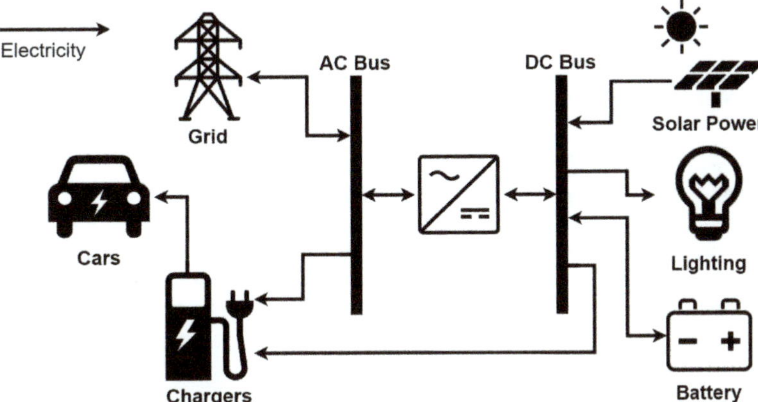

Figure 1. System configuration of a smart solar parking lot.

2.1. Solar Parking Lots

A solar array that was roof-mounted over the parking lot to generate electricity was considered. The total solar PV array generation capacity of 120 kWp was distributed over 40 parking spaces, corresponding to 3 kWp of generation per parking space. Power generation was simulated based on weather data from the Cabauw weather station located in the province of Utrecht in The Netherlands [13]. The data was used to simulate the typical power of the solar power array for one year with a time resolution of 15 min. Solar power generation was modeled using PVLib, a validated open-source tool developed by Sandia National Labs [14]. Table 1 describes the solar PV array characteristics used in the model.

Table 1. Description of solar photovoltaic array characteristics.

Characteristic	Value
Module technology	Monocrystalline silicon
Module rated power	300 kWp (60 cell)
Module rated efficiency	18.33% at STC
Array installed capacity	120 kWp
Site latitude	51°58' N
Site longitude	4°55' E
Array azimuth	0 °(South)
Array tilt	13 °
Parking spaces	40 spaces
Carport roof topology	Monopitch (single tilt angle for entire roof)
Annual production (DC)	133,625 kWh
Capacity factor (DC)	12.7%

2.2. Batteries

The electric vehicle batteries using the parking lot for charging were assumed to be representative of the current Dutch EV fleet, including BEVs and Plug-in Hybrid Electric Vehicles (PHEVs). The battery energy capacities considered therefore range from the 8.8 kWh Audi A3 PHEV to the 100 kWh Tesla Model X BEV. The solar parking lot also included a stationary Li-ion based battery storage system, for storing excess energy to further reduce the peak demand. The battery had a rated power of 50 kW and a capacity of 50 kWh, of which 80% was usable. The total number of batteries, N_b, is at all times less than 41, since the maximum occupancy of the parking lot is 40 EVs and there is always the stationary battery.

The charging efficiency, η_{chg}, and discharging efficiency, η_{dis}, in the battery, were each assumed to be 95%, leading to overall roundtrip losses of 9.75%. Coulomb counting was used to infer the state of charge (SoC) of the battery and changes to it. The rectification stage in the vehicle was assumed to lead to losses of about 6% in charging [15].

2.3. Electric Vehicle Supply Equipment

The system includes 40 charge points, each rated at 32 A (7.4 kW) for both AC and DC. While this rating is commonly found as AC level 2 charging [16], it is a lower current capacity than commercial DC charge points. The reason for this choice was to enable the slow charging of EV batteries on the DC bus without multiple rectification-inversion stages, as is expected in the future. The losses in the EVSE, which are primarily resistive in nature, were assumed to be around 0.2% [15].

2.4. Electric Vehicle Load Profile

The electric vehicle load profile was built based on two submodels—first, the EV arrival and departure model and the second, the estimation of the state of charge at the point of entry. In addition, the load is also determined by the extent to which the battery is to be charged by the time of departure. In this work, it is assumed that the EV drivers wish for their EVs to be charged to 100% SoC whenever possible.

2.4.1. EV Arrival and Departure

Direct Use of Observed Activity-Travel Schedule (DUOATS) models are a common method applied in smart charging and demand response studies [17]. Such a model was used in this work, whereby observed vehicle patterns were used to simulate EV behavior. The parking location considered was a workplace, and the model was based on data from the EV Project, a project run by the United States Department of Energy. It included data related to 8228 electric vehicles and hundreds of thousands of trips and charging events [18]. Since the data was collected from a large number of participants and geographical locations across the USA, it was assumed to be generalizable.

This historical parking data was used to determine the arrival and departure rates at both parking lots for each timestep of the day, considering weekdays and weekends separately. Based on these rates, a Monte Carlo approach was taken to determine the number of EV arrivals in each timestep. The duration of parking and the time of day were used to determine the number of departures in each timestep, after which an occupancy profile was built. A representative week of occupancy of vehicles at the workplace is shown in Figure 2. There are noticeable daily patterns of arrivals and departures during weekdays, with weekends having lower arrival rates.

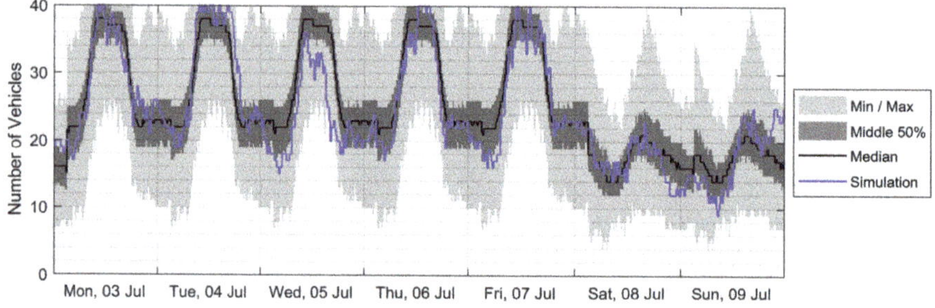

Figure 2. Simulated occupancy at the workplace parking lot over a week.

2.4.2. EV State of Charge on Arrival

Truncated normal distributions were used for assigning the SoCs on arrival. The coefficients for the normal distribution were inferred by fitting data from the EV Project [18], which collected data for over 8000 EVs in the USA. They are shown in Table 2.

Table 2. Coefficients describing the assignment of EV battery states of charge on arrival.

EV Type	Mean	Standard Deviation	Lower Bound	Upper Bound
BEV	50%	18%	0%	90%
PHEV	45%	30%	0%	90%

The lower mean SoC and greater standard deviation for PHEVs are explained by the lower concern of PHEV drivers about depleted batteries in comparison with BEV drivers.

Based on the time of arrival of an individual EV, the battery capacity of the EV and its SoC at the time of arrival, the expected charging demand of the EV was calculated. This demand (or as high a fraction as possible) needed to be met within the plug-in duration. The control system was informed about the time of departure at the time of arrival in all cases.

With knowledge of the available solar energy, the forecasted solar energy over the forecast horizon and the EV charging demand, the optimal charging of EVs in the parking space was to be determined in a manner that minimized the peak electricity demand in the solar parking lot. The following section describes the methods used for this scheduling.

3. Methods for Charge Scheduling

For an investigation into the effect of uncertainty of EV demand on peak loads, two reference cases and three scenarios are considered. One reference case is unscheduled charging, where EVs charge at maximum power as soon as they are plugged in. In addition, a case is simulated with perfect forecasting of solar production and EV demand, which may be considered as another reference case. The three scenarios investigated lie between these two extreme cases:

1. No EV demand forecast: EV charging is scheduled without a forecast of energy demand for EVs arriving in the future
2. Average EV demand forecast: EV charging is scheduled with a single forecast of energy demand for EVs arriving in the future which is based on average values.
3. Robust EV demand forecast: EV charging is scheduled to be robust across a range of possible energy demands for EVs arriving in the future

In all these three scenarios, the schedule was designed to be robust across a range of possible solar forecasts. These scenarios thus differ only in their approach to EV demand forecasting. The objective function for peak reduction under perfectly accurate forecasts is described in Section 3.1. The introduction of uncertainty in the forecasts is described in Section 3.2, after which each of the three scenarios and its optimization formulations are described. The commercial solver, Gurobi, was used with MATLAB in each case on a Windows PC with an Intel i5 1600 MHz quad-core processor and 32GB RAM.

3.1. Problem Formulation with Perfect Forecasting

In the scenario with perfect forecasting, the future solar production, as well as future electric vehicle demand over the 24 h horizon in the future, are assumed to be accurately known in advance. This is not a practically feasible scenario since neither of these can be accurately known in advance. However, this scenario clearly defines the best possible performance of the scheduling approach, with reference to which other scenarios may be compared. In addition, it also provides an idea of the performance of the scheduling algorithm independent of the degree of accuracy of the forecast.

To optimize the scheduling of EVs within the carport, we apply Model Predictive Control (MPC). MPC is a control technique used for determining the optimal behavior of complex multivariate problems. The control action is determined at each time step by solving an open-loop optimal control problem over a finite time horizon. MPC is used to solve problems during the operation of the system, taking the state of the system into account at each time step. These characteristics make MPC suitable for the control of EV charge scheduling at the solar parking lot.

The goal of the scheduling of EV charging is to reduce the peak demand of the solar parking lot over the time horizon under consideration. Thus, the objective function is:

$$\text{minimize} \quad \max\left(E_{grid}(k), \ldots, E_{grid}(k + N_p - 1)\right), \tag{1}$$

where $E_{grid}(k)$ is the net energy exchange between the parking lot and the grid at time k and N_p is the time horizon, which is 24 h. A sensitivity analysis on the duration of the horizon revealed that longer forecasts had no increased benefits to the simulation. Further, with reducing the accuracy of the length of the simulation, the reliability of longer horizon forecasts is lower upon the introduction of forecasting uncertainty. An auxiliary variable, $E_{grid}^{max}(t)$ is introduced, which represents the local maxima or peak in grid exchange of the parking lot with N_b batteries and lighting load, E_{load} over the considered horizon, N_p.

The objective function is thus rewritten as:

$$\underset{E_{grid}^{max}, E, \delta, z}{\text{minimize}} \quad E_{grid}^{max} \tag{2}$$

with the decision variables

$$E_{grid}^{max}, E_i(t), \delta_i(t), z_i(t) \quad \text{for } i \in \{1, \ldots, N_b\}, \quad t \in \{k, \ldots, k + N_p - 1\}, \tag{3}$$

subject to a number of constraints, described below.

At any given time, t, the energy exchanged with the grid, $E_{grid}(t)$, depends on the PV production, $E_{PV}(t)$, the lighting load, $E_{load}(t)$, and summed load of each battery, $E(t)$. It is ensured that the energy exchange peak, which is subject to minimization, is the highest peak within the considered horizon both in purchase as

$$E_{grid}^{max} \geq E_{load}(t) - E_{PV}(t) + \sum_{i=1}^{N_b} E_i(t) \ \forall t, \tag{4}$$

as well as in feed in as

$$E_{grid}^{min} \leq E_{load}(t) - E_{PV}(t) + \sum_{i=1}^{N_b} E_i(t) \ \forall t. \tag{5}$$

The stored energy in the ith battery in the $(k+1)^{th}$ timestep, the state variable, $S_i(k+1)$, is described as

$$S_i(k+1) = \begin{cases} S_i(k) + \eta_{chg,i} \cdot E_i(k), & E_i(k) > 0 \\ S_i(k) + \frac{1}{\eta_{dis,i}} E_i(k), & E_i(k) \leq 0 \end{cases}. \tag{6}$$

In order to formulate the battery behavior linearly, the Mixed Logical Dynamics (MLD) formalism is used [19]. A binary decision variable $\delta_i(t)$ is introduced, defined as:

$$[\delta_i(t) = 0] \leftrightarrow [E_i(t) > 0] \ (\text{EV or battery is charging}) \tag{7}$$

$$[\delta_i(t) = 1] \leftrightarrow [E_i(t) \leq 0] \ (\text{battery is discharging}), \tag{8}$$

and leads to the constraint:

$$\delta_i(t) \in \{0,1\} \ \forall \ i, \ t. \tag{9}$$

This binary decision variable, $\delta_i(t)$, is then used to reformulate Equation (6) as

$$S_i(k+1) = S_i(k) + \eta_{chg,i} \cdot E_i(k) \cdot (1 - \delta_i(k)) + \frac{1}{\eta_{dis,i}} E_i(k) \cdot \delta_i(k). \tag{10}$$

However, this formulation is still nonlinear because it contains the product of two decision variables, $E_i(k)$ and $\delta_i(k)$. An additional set of continuous decision variables is introduced as

$$z_i(t) = \delta_i(t) \cdot E_i(t). \tag{11}$$

The minimum energy, which can be stored in the ith, battery in the $(t+1)^{th}$ timestep, is limited by the lowest possible energy which can be delivered to it with the objective of maximimizing the SoC at departure. This is formulated as

$$S_i^{\min}(t+1) \ \leq S_i(t) + \eta_{chg,i}E_i(t) + \left(\frac{1}{\eta_{dis,i}} - \eta_{chg,i} \right) z_i(t) \ \forall i, \ t. \tag{12}$$

The maximum energy which can be stored in the i-th, battery in the $(t+1)$-th timestep is limited by the physical constraints on the battery capacity and the power rating of the chargepoints. It is formulated as

$$S_i^{\max}(t+1) \ \geq S_i(t) + \eta_{chg,i}E_i(t) + \left(\frac{1}{\eta_{dis,i}} - \eta_{chg,i} \right) z_i(t) \ \forall i, \ t. \tag{13}$$

The constraints on the energy exchanged with the ith battery in each time step is given as

$$E_i(t) \leq M_i \cdot (1 - \delta_i(t)) \qquad\qquad \forall i, t \tag{14}$$
$$E_i(t) \geq \varepsilon + (m_i - \varepsilon) \cdot \delta_i(t) \qquad\qquad \forall i, t, \tag{15}$$

where M_i is the maximum allowable value of $E_i(t)$ and m_i is the minimum allowable value. An overestimate of M_i or an underestimate of m_i is acceptable, but values close to the true maximum and minimum are preferred to lower computational time. These values are taken as

$$M_i = P_i^{max} \cdot \Delta t \tag{16}$$
$$m_i = -P_i^{max} \cdot \Delta t, \tag{17}$$

where P_i^{max} and P_i^{min} are the power limits of the ith battery and Δt is the length of the time step i.e., 15 min. The tolerance, ε is a small value, typically the machine precision of the solver.

The following constraints then limit the auxiliary MLD variables $\delta_i(t)$ and $z_i(t)$ to ensure they will be equivalent to their stated definitions [20]

$$z_i(t) \leq M_i \cdot \delta_i(t) \qquad\qquad \forall i, t \tag{18}$$
$$z_i(t) \geq m_i \cdot \delta_i(t) \qquad\qquad \forall i, t \tag{19}$$
$$z_i(t) \leq E_i(t) + M_i \cdot (1 - \delta_i(t)) \qquad\qquad \forall i, t \tag{20}$$
$$z_i(t) \geq E_i(t) + m_i \cdot (1 - \delta_i(t)) \qquad\qquad \forall i, t. \tag{21}$$

3.2. Inclusion of Uncertainty in Forecasting

Since historic data of the PV yield and modeled data for the EV demand were used, the values of the future PV yield and EV demand were known. In order to simulate an inaccurate forecast, errors were introduced to the known PV production and the EV demand over the relevant horizon.

3.2.1. Uncertainty in PV Forecasting

$E_{PV}(t)$ is known to be the PV production over the relevant horizon at the timestep, t. The forecasted value of PV production is the sum of $E_{PV}(t)$ and an additional solar forecasting error term, $\omega_{PV}(t)$, as

$$E_{fcst}(t) = E_{PV}(t) + \omega_{PV}(t). \tag{22}$$

Monte Carlo methods are used in this case to generate a finite but large number of error vectors, Ω^*_{PV}, which are then considered.

$$\Omega^*_{PV} = \{\omega^{(1)}_{PV}, \ldots, \omega^{(N_e)}_{PV}\} \subseteq \Omega_{PV}, \tag{23}$$

where N_e is the number of error vectors considered. If N_e is large enough, Ω^*_{PV} may be considered to be a reasonably good approximation of Ω_{PV}, the set of all possible error vectors. 10,000 is chosen for N_e in this case. Though forecasting error is normally distributed, for robust optimization, a bounded distribution is required. The distribution is therefore truncated such that

$$-3\sigma_{PV}(t) \leq \omega^j_{PV} \leq 3\sigma_{PV}(t), \tag{24}$$

for all t in $k, \ldots, k + N_P - 1$ and j in $1, \ldots, N_e$. The upper and lower bounds are determined by $\sigma_{PV}(t)$ the standard deviation of the forecasting error at time, t. The forecasting error is also truncated so that the forecasted power generation cannot be less than zero or greater than the clear sky generation. The choice of three standard deviations as a limit, rather than more conservative values of five or seven, is justified based on simulation results.

3.2.2. Uncertainty in EV Forecasting

Two approaches were considered here—the average approach and the Monte Carlo approach, as used in the PV uncertainty introduction. In the average approach, the terms $S_i^{\min}$ and $S_i^{\max}$, defined in Equations (12) and (13) are taken to be their average values based on data collected from EVs in the parking lot at the timestep, t. Thus, uncertainty in a number of variables like arrival time and numbers, departure times, the energy capacity of the vehicle, the SoC of the vehicle on arrival are all clustered together to be dealt with through the optimization formulation. The nature of the formulation implies a single forecast is available in each timestep, which is based upon average values in the past, thus satisfying the aim of this method of introduction of uncertainty.

In the second approach, error terms are introduced through Monte Carlo simulation, as with the errors in PV generation forecasts. However, as in the average case, the errors are introduced in the terms $S_i^{\min}$ and $S_i^{\max}$, influencing the state constraints rather than the state variables. This formulation thus considers a range of possible forecasts, treated as equally probable, over which the problem needs to be solved.

3.3. No EV Demand Forecast

No EV forecasting is the simplest strategy where the charging of EVs plugged-in at the parking lot at the current time step are optimized over the period they are expected to be plugged in. The arrival of additional EVs in the near future and their demands are not considered—a drawback of the approach.

The schedule is designed to deal with uncertainty in the PV forecast. It does this by considering a range of possible PV forecasts, over all of which it reduces peaks. In other words, it operates robustly over a range of PV uncertainties.

To ensure robustness, we use min max optimization, which minimizes the cost function over the decision variables for the worst case i.e., highest peak load. In this case, the objective function is

$$\underset{E_{grid}^{max}, E_i(t), \delta_i(t), z_i(t)}{\text{minimize}} \quad \underset{\omega \in \Omega_{PV}(t)}{\text{maximize}} \quad E_{grid}^{max}, \tag{25}$$

where ω is a vector representing a random possible value for the forecasting errors at each time step t, and $\Omega_{PV}(t)$ is the bounded set of all possible forecasting errors. The errors at each time step are therefore drawn randomly from the uniform distribution given by

$$\omega^{j}_{PV}(t) \in \mathcal{U}(-3\sigma_{PV}(t), 3\sigma_{PV}(t)). \tag{26}$$

A new auxiliary variable, T, is now defined as the maximum value for the objective function under all the forecasting errors considered. Because the objective function is not directly dependant on the forecasting uncertainty, T can simply be defined as being equal to the original objective function. If J is the maximum peak in each case of forecasting error considered, and $\tilde{u}$ the decision variables, T is given by

$$T = \max_{\tilde{u}}(J(\omega^{(1)}_{PV}), \ldots, J(\omega^{(N_e)}_{PV})) = E^{max}_{grid}, \tag{27}$$

where E^{max}_{grid} is independent of the uncertainty variables.

The objective function is redefined as

$$\underset{T, E^{max}_{grid}, E, \delta, z}{\text{minimize}} \quad T, \tag{28}$$

with the decision variables:

$$T\, E_i(t),\ \delta_i(t),\ z_i(t) \qquad \text{for } i \in \{1, \ldots, N_b\}, \quad t \in \{k, \ldots, k + N_p - 1\}. \tag{29}$$

The constraints in Equations (4) and (5) are changed to include the solar forecasting error as

$$E^{max}_{grid} \geq E_{load}(t) - E_{fcst}(t) + \omega^{max}_{PV}(t) + \sum_{i=1}^{N_b} E_i(t) \qquad \forall\, i,\, t \tag{30}$$

$$E^{min}_{grid} \leq E_{load}(t) - E_{fcst}(t) + \omega^{min}_{PV}(t) + \sum_{i=1}^{N_b} E_i(t) \qquad \forall\, i,\, t \tag{31}$$

The optimization problem is then solved to be robust across the errors in PV forecast, $\omega_{PV}(t)$, which lie within the bounded range of uncertainty $\Omega_{PV}(t)$. As proven in [20], a constraint cannot be active at some intermediate value of the disturbance without violating the constraint at the extreme value. In this model, the greater-than constraint will be active only at the maximum value of the disturbance and a less-than constraint will be active only at the minimum. Hence, only the minimum value, $\omega^{min}_{PV}(t)$, and maximum value, $\omega^{max}_{PV}(t)$, of the errors were considered, which are sufficient for the all intermediate error values.

Although there is no EV forecasting, vehicles which are not plugged in at the charging station at the relevant timestep, would still begin charging at some point in the future. These limits are then included in constraints in Equations (12) and (13), which are modified to

$$S^{min}_i(t+1) \begin{cases} = 0, \text{ when a vehicle is not present at the current timestep, } t, \text{ in space, } i \\ \leq S_i(t) + \eta_{chg,i} E_i(t) + \left(\frac{1}{\eta_{dis,i}} - \eta_{chg,i}\right) z_i(t) \ \forall\, i,\, t \\ \text{when a vehicle is present at the current timestep, } t, \text{ in space, } i \end{cases} \tag{32}$$

and

$$S^{max}_i(t) \begin{cases} = 0, \text{ when a vehicle is not present at the current timestep, } t, \text{ in space, } i \\ \geq S_i(t) + \eta_{chg,i} E_i(t) + \left(\frac{1}{\eta_{dis,i}} - \eta_{chg,i}\right) z_i(t) \ \forall\, i,\, t \\ \text{when a vehicle is present at the current timestep, } t, \text{ in space, } i \end{cases} \tag{33}$$

The constraints in Equation (15) through Equation (21) remain unchanged.

3.4. Average EV Demand Forecast

The objective function with robust solar PV forecast and a single average EV forecast remains the same as Equation (28). The stored energy terms, $S_i^{\min}(t)$ and $S_i^{\max}(t)$ in the constraints in Equations (12) and (13) are replaced by average values of these variables. The new constraints are therefore:

$$\overline{S_i^{\min}(t+1)} \leq \overline{S_i(t)} + \eta_{chg,i} E_i(t) + \left(\frac{1}{\eta_{dis,i}} - \eta_{chg,i} \right) z_i(t) \qquad \forall\, i,\, t \qquad (34)$$

$$\overline{S_i^{\max}(t+1)} \geq \overline{S_i(t)} + \eta_{chg,i} E_i(t) + \left(\frac{1}{\eta_{dis,i}} - \eta_{chg,i} \right) z_i(t) \qquad \forall\, i,\, t, \qquad (35)$$

where $\overline{S_i^{\min}(t)}$ and $\overline{S_i^{\max}(t)}$ are the average values for the variables $S_i^{\min}(t)$ and $S_i^{\max}(t)$ respectively.

3.5. Robust EV Demand Forecast

The system is meant to be robust in the sense of reducing peak grid exchange across a wide range of EV forecasting errors as well as errors in the solar forecast. The maximum value, $\omega_i^{\max}(t)$ and minimum value, $\omega_i^{\min}(t)$, in the range of errors introduced in the EV demand forecast through the Monte Carlo simulation, as described in Section 3.2.2, were used to define the state constraints for $S_i^{\min}(t)$ and $S_i^{\max}(t)$ as

$$S_i^{\min}(t) + \omega_i^{\max}(t) \leq S_i(t) + \eta_{chg,i} E_i(t) + \left(\frac{1}{\eta_{dis,i}} - \eta_{chg,i} \right) z_i(t) \qquad (36)$$

$$S_i^{\min}(t) + \omega_i^{\min}(t) \geq S_i(t) + \eta_{chg,i} E_i(t) + \left(\frac{1}{\eta_{dis,i}} - \eta_{chg,i} \right) z_i(t). \qquad (37)$$

However, constraints risk making the problem infeasible if $\omega_i^{\max}(t) > \omega_i^{\min}(t)$, which is highly likely. In order to enable a feasible solution to the problem, the state constraints for vehicles which are forecasted to arrive, are expressed as soft constraints:

$$S_i^{\min}(t) + \omega_i^{\max}(t) \leq S_i(t) + \eta_{chg,i} E_i(t) + \left(\frac{1}{\eta_{dis,i}} - \eta_{chg,i} \right) z_i(t) + \epsilon_i^1(t) \qquad (38)$$

$$S_i^{\min}(t) + \omega_i^{\min}(t) \geq S_i(t) + \eta_{chg,i} E_i(t) + \left(\frac{1}{\eta_{dis,i}} - \eta_{chg,i} \right) z_i(t) - \epsilon_i^2(t). \qquad (39)$$

for all $i \in \{1, \ldots, N_b - 1\}$, $t \in \{k, \ldots, k + N_p - 1\}$. The slack variables $\epsilon_i^1(t) \geq 0$ and $\epsilon_i^2(t) \geq 0$ correspond respectively to the constraints for the minimum and maximum stored energy, and are added to the optimization problem as an auxiliary decision variables. For the stationary battery as well as the vehicles plugged in at the parking lot at a given timestep, k, the state constraints remain hard since the minimum and maximum stored energies are known with certainty.

The complete optimization formulation is then given by:

$$\underset{T, E_{grid}^{\max}, E, \delta, z, \epsilon}{\text{minimize}} \quad T + \frac{1}{(N_b - 1) \cdot N_p} \sum_{i=1}^{N_b - 1} \sum_{t=k}^{k + N_p - 1} c_1 \epsilon_i^1(t) + c_2 \epsilon_i^2(t), \qquad (40)$$

where c_1 and c_2 are penalty constants. The values $c_1 = 1$ and $c_2 = 1$ were found empirically to lead to lowest values of peak demand. Increasing the penalty constant values did not however lead to large changes in peak demand. The decision variables are:

$$T,\, E_{grid}^{\max},\, E_i(t),\, \delta_i(t),\, z_i(t),\, \epsilon_i^1(t), \epsilon_i^2(t) \qquad \text{for } i \in \{1, \ldots, N_b\}, \quad t \in \{k, \ldots, k + N_p - 1\}, \qquad (41)$$

with the additional constraint

$$\epsilon_i^1(t), \, \epsilon_i^2(t) \, \geq 0 \ \ \forall \, i, \, t. \tag{42}$$

4. Results and Discussion

A year of operation of the solar parking lot was simulated for each of the three described scenarios to determine the performance of the system at peak reduction in each case. During the comparison, the reference cases of unscheduled charging and perfect forecasts are also included to provide additional insights.

4.1. Example Simulations

Figure 3 shows the energy flows within the system over a representative period, in this case, a week. The two reference cases, unscheduled charging and charging with perfect forecasting, are compared. The characteristics of the scheduling seen here are also valid for longer simulations over the year.

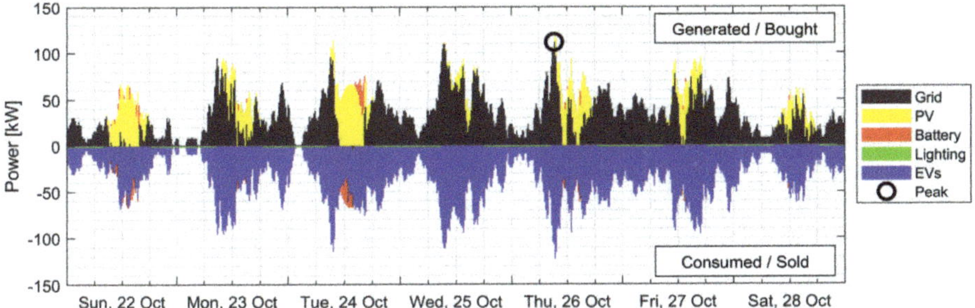

(a) Energy flows (unscheduled charging).

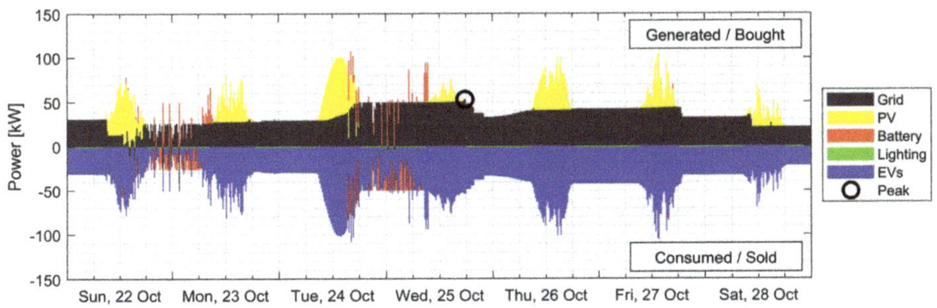

(b) Energy flows (perfect forecasting).

Figure 3. Comparison of energy flows in the solar parking lot with unscheduled charging as opposed to charging with perfect forecasts.

In accordance with thermodynamic system conventions, the energy entering the electrical system (PV production, grid purchase and battery discharge) is taken as positive and energy leaving the system (EV and battery charging, lighting load and feed-in to the grid) as negative. The shape of the positive and negative sides are similar, showing the energy balance in each timestep, excluding losses. The solar energy production is shown in yellow, battery charge and discharge in red and EV charging in blue. The lighting loads, which are very small in comparison with the others, are shown in green and residual grid load (after solar self-consumption and battery charge/discharge) in gray. The highest peak in the week is highlighted in each figure.

The largest peak in the residual grid load (110 kW) as a result of unscheduled charging, seen on Thursday in Figure 3a, is considerably reduced in magnitude to 50 kW as a result of scheduled charging with perfect forecasts. In addition, the frequency of these peaks is found to decrease considerably and a uniformly flat load profile achieved. The arrivals of many EVs on Tuesday, Wednesday and Thursday mornings, all of which lead to peaks in energy demand if unscheduled are all adequately shaved. Although the perfect forecasting scenario is not an applicable case, it demonstrates the success of the method at reducing the peak residual load in the week considered.

Figure 4, similarly, shows the energy flows in the system for the three scenarios. As seen in Figure 4a, where no EV forecast is available, the magnitude of the highest peak in this case (100 kW) is lower than that in the unscheduled case. Relative to the perfectly forecasted case, though, the peaks are considerably higher. A drawback of the scenario can be seen in the situation leading to the Wednesday afternoon peak. The EVs in the parking lot on Tuesday night was charged slowly to reduce the peak load at night. This led to a larger number of vehicles in need of simultaneous charging on Wednesday morning, leading to a high peak in electricity demand in the afternoon. On the other hand, the Thursday peak seen in the unscheduled charging case was effectively shifted.

The single average forecast seen in Figure 4b already provides a considerable reduction in the magnitude of the highest peak (85 kW) relative to the case without an EV forecast. On the same Wednesday, the vehicles parked overnight are charged in the morning in anticipation of future arrivals, thus lowering the peak on Wednesday. However, a drawback of the system is seen on Monday, which has a lower than average EV charging demand. A large number of EVs are charged in the morning in anticipation of demand later in the day. However, the demand in the afternoon was lower than expected, making the morning peak unnecessary in hindsight. This leads to Monday having the highest peak in the week.

The robust treatment of forecasting seen in Figure 4c results in even further reduction in the magnitude of the highest peak in the week, which is about 75 kW. The Wednesday load is lowered even more, and there is a better performance in the case of the lower demand on Monday. Similar to the average forecast, there are many peaks with similar magnitude across the week, but the height of the highest one is lower than in the case without the demand forecast.

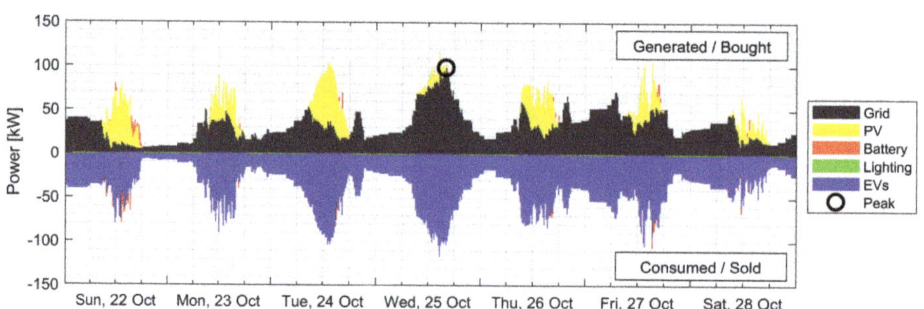

(**a**) Energy flows (no electric vehicle (EV) demand forecast).

Figure 4. *Cont.*

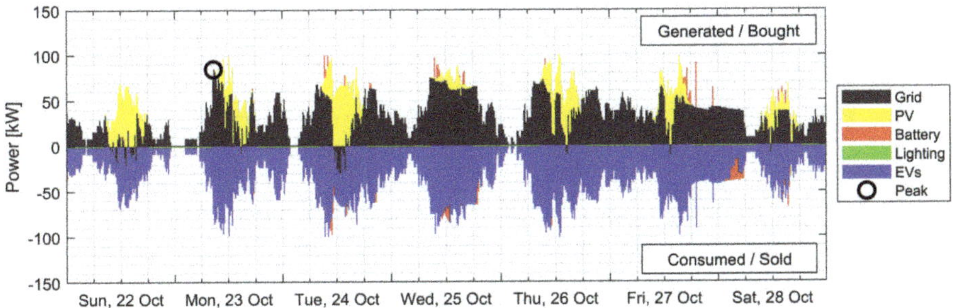

(**b**) Energy flows (average EV demand forecast).

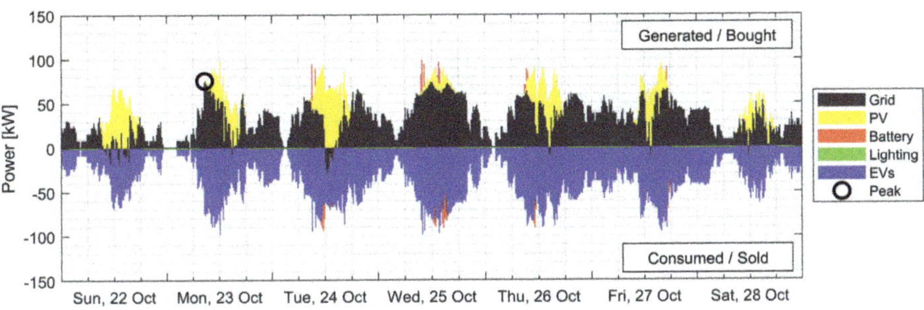

(**c**) Energy flows (robust EV demand forecast).

Figure 4. Comparison of energy flows in the solar parking lot with no EV demand forecast, a single average EV demand forecast and a robust consideration of EV demand forecasting.

It can be seen that in the system, the peak loads on the grid are always the result of EV charging and never the result of solar feed-in. While solar peaks did occur under low occupancy of the parking lot during summer when the stationary battery was full, they were generally lower than the EV charging peaks. Further, it was assumed that curtailment would be the strategy for peak shaving of solar feed-in peaks.

The power flows also reveal that the use of fixed storage remains low in all cases. A sensitivity analysis was conducted on the battery size used. It revealed that small batteries had a considerable impact on peak reduction, but the returns diminished with increasing battery size. Increasing the battery size beyond 50 kWh had no further effect on peak reduction.

4.2. Maximum Annual Peak Exchange with the Grid

The scenarios are compared in terms of a few metrics, which provide insight into the simulation results. The first metric considered is the highest annual peak in each scenario. These peaks have a magnitude which is generally close to the transformer rated capacity and occur rarely—a few times in the transformer lifetime. They lead to a type of transformer loading known as short term emergency loading, involving a very high demand occurring for periods of half an hour or less. However, despite the short duration and relative rarity of their occurrence, these overloads can cause considerable damage. Increased hot-spot temperatures, resulting in the evolution of free gas from insulation and insulating fluid, reduced mechanical strength and deformation of conductors and structural insulation, and high internal pressures resulting in leaking gaskets or loss of oil can all be results of short term emergency loading. These loads can considerably shorten the lifespan of these assets [21]. A reduction in the highest annual peak indicates a reduction in the intensity of these events. Figure 5 illustrates the peak annual power demand compared across the three scenarios and the two reference cases.

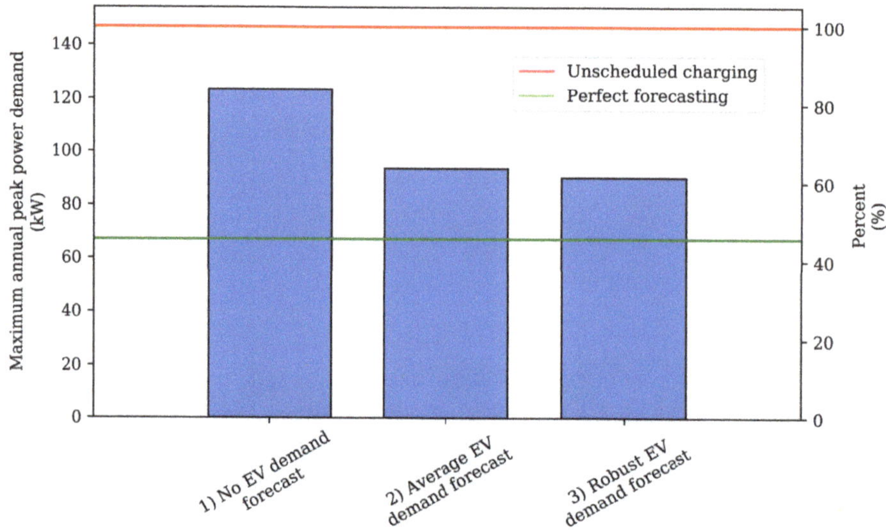

Figure 5. Comparison of annual peak power exchange with the grid.

The comparison shows that smart charging for peak reduction in the absence of EV demand forecasting is effective at the reduction of short term peak loads, but this effect is limited (about 16% peak reduction relative to unscheduled charging). Further, the magnitude of peak reduction in the absence of EV demand forecasting is considerably less than that possible with perfect forecasting of future EV demand (about 54% reduction), which is the case often assumed in the literature. The availability of single forecasts based on average demands in the past result in an increase in the effectiveness of reducing short term peak loads. Peaks are reduced by an additional 20% relative to unscheduled charging. Consideration of multiple possible forecasts across which the system works in a robust manner further reduces the peak power exchange of the system. Such a system had an annual peak of 39% lower than that found in unscheduled charging.

4.3. Duration of Peak Loads

While the previous metric considered short term load intensity, it did not consider the frequency of occurrence of peaks of marginally smaller magnitude, whose impact is similar to that of the annual peak. A comparison of load duration curves can provide further insight. Figure 6 focuses on the leftmost section of the curve, where high peaks are visualized:

The downward shift of the y-intercept through smart charging represents the reduction in the magnitude of annual peak loads, whereas the leftward shift near the y-intercept represents the reduction in the number of peaks. Smart charging without EV demand forecasting, in addition to lowering the magnitude of peaks, is also found to reduce the number of peaks. The curve reveals that the times for which loads are greater than 100 kW are reduced by more than half while the times for which loads are greater than 80 kW are reduced by about half.

The provision of EV demand forecasting further reduces these durations. Loads above 100 kW are avoided altogether and the number of peaks greater than 80kW are greatly reduced. There is no clear advantage of robust forecasting over average forecasting in this case, with both methods providing similar reductions in the number of peaks that the asset is exposed to.

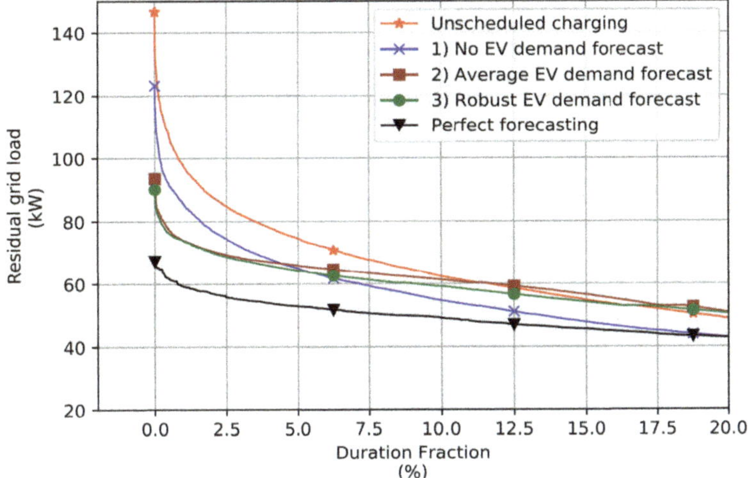

Figure 6. Load duration curves.

5. Conclusions

The goal of this work was to quantify the peak load increase when uncertainty is involved in charge scheduling of electric vehicles at a solar parking lot. It further aimed to develop strategies for scheduling charging in a manner that minimized the peak electricity load at the point of common coupling of the parking lot while taking this uncertainty into account. Since short duration high peaks have the maximum impact on transformer aging, these were the peaks that were focused on.

The set up considered included a solar parking lot with 40 spaces located at a workplace. It included a 120 kWp solar array, 40 EV charge points and a 50 kWh stationary battery. The arrival and departure of EVs, which were parked and plugged in at the parking lot, were simulated over a year. Model Predictive Control (MPC) was the method used to optimally schedule the charging of EVs in the parking lot over the year. The operation of the system was simulated over a year in terms of the energy exchanged by the parking lot with the grid.

The system was considered in three scenarios:

1. No EV demand forecast: EV charging is scheduled without a forecast of energy demand for EVs arriving in the future
2. Average EV demand forecast: EV charging is scheduled with a single forecast of energy demand for EVs arriving in the future which is based on average values.
3. Robust EV demand forecast: EV charging is scheduled to be robust across a range of possible energy demands for EVs arriving in the future

with a and a schedule that was robust across multiple possible EV demand forecasts. The scheduling for each scenario was formulated as an optimization problem. The operation of the solar carport was simulated in each scenario for a year based on the solution of the optimization problem. The scenarios were compared with two reference cases—unscheduled charging, which is the current norm, and charging with perfect forecasting of EV demand, which represents the limits of the effectiveness of the system at peak reduction.

The results show that for parking locations with charging, which are currently close to peak load capacity, scheduling of EVs can be used to reduce both the magnitude as well as the frequencies of peak loading on distribution level assets. The magnitude of the peak reduction is however considerably less than the peak reduction possible with perfect forecasting of future EV demand, which is often considered in the literature. Table 3 displays the results of annual peak reduction in the scenarios considered:

Table 3. Annual peak power across scenarios.

Nr.	Scenario	Annual Peak Power (kW)	Relative Peak Reduction (%)
Ref	Unscheduled charging	147	0%
1	No EV forecast	123	16% ($\downarrow$)
2	Average EV forecast	94	36% ($\downarrow$)
3	Robust EV forecast	90	39% ($\downarrow$)
Ref	Perfect forecasting	67	54% ($\downarrow$)

Without EV demand forecasting, the maximum annual peak load of the solar carport was reduced by 16% in our case relative to unscheduled charging. This was, however, considerably less effective than in the reference case with perfect forecasting, where the magnitude of the annual peak was reduced by 54%. The inclusion of a single 24 h horizon EV forecast reduced the peak in the solar parking lot by 36%, increasing the effectiveness of the scheduled charging by an additional 20%. Consideration of multiple forecasts of possible EV demand and robust adjustment of the schedule for the performance of the worst possible forecast marginally improved the effectiveness of the scheduling, reducing the peak by 39%.

In addition to reducing the magnitude of peak loads, scheduling of EV charging also has the effect of reducing the number of peaks that distribution level assets were subject to. The use of EV demand forecasting was found to have the effect of considerably reducing this number. However, in this case, the consideration of multiple forecasts provides no clear advantage over a single forecast.

An economic analysis of the system was considered out of the scope of this work. As such, the cost-benefit analysis of scheduling EV charging versus upgrade of the grid connection was not performed. However, preliminary investigation indicates that there is considerable value for the parking lot owner through the implementation of the system described in this work. In the USA, capacity charges for the grid connection at EV charging sites can be higher than $2000/month, causing the electricity utility bills of some businesses to increase by a factor of four [22]. Similarly, in the Netherlands, the grid capacity cost is € 190/year per charge point or 37% of the annual operational costs for the charge point excluding energy costs and about 20% of the costs including energy [23]. Although case-specific, peak reduction does have considerable economic value for system operators. This is expected to be addressed in future work.

Additionally, future research work also involves the investigation of the dependence of the scheduling on the location of the parking lot i.e., on whether the parking patterns have an influence on the choice of the objective of the charging schedule. The use of scheduling for off-grid or constrained grid capacity design of longterm parking lots for EVs may be considered. Improved methods for EV scheduling for other objectives will also be addressed in future works.

Author Contributions: R.G. wrote the final draft, performed some of the analysis and managed the project. Y.S. conceptualized the methodology, performed the investigation and wrote the first draft. S.F. provided guidance with the formulation of the optimization problems and their solution. Z.L. provided supervision and edited the final draft. A.v.W. provided supervision, edited the final draft and acquired funding for this work. All authors have read and agreed to the published version of the manuscript.

Funding: This research was funded by the European Funds for Regional Development through the Kansen voor West program grant 00113 for the project, Powerparking, as well as the Netherlands Organization for Scientific Research as part of the Uncertainty Reduction in Smart Energy Systems (URSES+) grant for the project 'Car as Power Plant-LIFE'.

Acknowledgments: The authors would like to thank Tomás Pippia for helpful discussions that contributed to this work.

Conflicts of Interest: The authors declare no conflict of interest.

List of Symbols

Symbol	Definition	Unit	Note
$E_{fcst}(t)$	Forecasted PV generation: $E_{PV}(t) + \omega_{PV}(t)$	kWh	
$E_{grid}^{\max}$	$\max\left(E_{grid}(k), \ldots, E_{grid}(k+N_p-1)\right)$	kWh	
$E_{grid}^{\min}$	Max energy that can be sent to the grid	kWh	32 kWh = 120 kW $\cdot 1.05 \cdot \Delta t$
$E_{grid}(t)$	Grid exchange: $E_{load}(t) - E_{PV}(t) + \sum_{i=1}^{N_b} E_i(t)$	kWh	
$E_i(t)$	Energy to (+) or from (-) battery i at time t	kWh	
$E_{load}(t)$	Load from lighting at time t	kWh	
$E_{PV}(t)$	Generation from solar power at time t	kWh	
i	Index for each battery, 1–40 = EVs, 41 = fixed storage	-	$i \in \{1, \ldots, N_b\}$
k	Current time step	-	$k \in \{1, \ldots, N_T\}$
M_i	Max possible value of $E_i = P_i^{\max} \cdot \Delta t$	kWh	
m_i	Min possible value of E_i: $= -P_i^{\max} \cdot \Delta t$	kWh	
N_b	Total number of batteries	-	41 = 40 EVs + 1 battery
N_e	Number of errors in the bounded set	-	10,000
N_p	Number of time steps in MPC time horizon	-	$96 = 24 \cdot 4$
N_T	Number of time steps in one full simulation	-	$34{,}944 = 24 \cdot 4 \cdot 364$
$P_i^{\max}$	Maximum power to or from battery i	kW	EVs 7.4 kW, battery 50 kW
$S_i(t)$	Energy stored in battery i at time t	kWh	
$S_i^{\max}(t)$	Maximum energy allowed in battery i at time t	kWh	
$S_i^{\min}(t)$	Minimum energy allowed in battery i at time t	kWh	
$\overline{S_i^{\max}}(t)$	Average value for the max energy in battery i at time t	kWh	
$\overline{S_i^{\min}}(t)$	Average value for the min energy in battery i at time t	kWh	
t	time step within MPC horizon	-	$t \in \{k, \ldots, k+N_p-1\}$
$z_i(t)$	$z_i(t) = \delta_i(t) \cdot E_i(t)$	kWh	
Δt	Length of a single time step	h	15 min = 0.25 h
$\delta_i(t)$	For battery i at time t: 0 if discharging, 1 if charging	$\{0,1\}$	
$\eta_{chg,i}$	Charging efficiency of battery i	-	
$\eta_{dis,i}$	Discharging efficiency of battery i	-	
$\Omega_{PV}^*(t)$	Bounded set of PV forecasting errors $\{\omega_{PV}^{(1)}, \ldots, \omega_{PV}^{(N_e)}\}$	-	
$\omega_{PV}(t)$	PV forecasting error at time t	kWh	
$\omega_{PV}^{\max}(t)$	Max PV forecasting error in set $\Omega_{PV}^*(t)$	kWh	
$\omega_{PV}^{\min}(t)$	Min PV forecasting error in the set $\Omega_{PV}^*(t)$	kWh	
$\omega_{S_i^{\max}}(t)$	Uncertainty in the value of $S_i^{\max}(t) - S_i^{\min}(t)$	kWh	
$\omega_{S_i^{\min}}(t)$	Uncertainty in the value of $S_i^{\min}(t) - S_i^{\min}(t-1)$	kWh	

References

1. Olivier, J.; Schure, K.; Peters, J. *Trends in Global CO$_2$ and Total Greenhouse Gas Emissions—2017 Report*; Technical Report 2674; PBL Netherlands Environmental Assessment Agency: The Hague, The Netherlands, 2017.
2. Van Mierlo, J.; Messagie, M.; Rangaraju, S. Comparative environmental assessment of alternative fueled vehicles using a life cycle assessment. *Transp. Res. Procedia* **2017**, *25*, 3435–3445. [CrossRef]
3. van Vliet, O.; Brouwer, A.S.; Kuramochi, T.; van den Broek, M.; Faaij, A. Energy use, cost and CO$_2$ emissions of electric cars. *J. Power Sources* **2011**, *196*, 2298–2310. [CrossRef]
4. Dubey, A.; Santoso, S. Electric Vehicle Charging on Residential Distribution Systems: Impacts and Mitigations. *IEEE Access* **2015**, *3*, 1871–1893. [CrossRef]
5. Verzijlbergh, R.A.; Lukszo, Z.; Slootweg, J.G.; Ilic, M.D. The impact of controlled electric vehicle charging on residential low voltage networks. In Proceedings of the 2011 International Conference on Networking, Sensing and Control, Delft, The Netherlands, 11–13 April 2011; pp. 14–19. [CrossRef]
6. Uddin, M.; Romlie, M.F.; Abdullah, M.F.; Abd Halim, S.; Abu Bakar, A.H.; Chia Kwang, T. A review on peak load shaving strategies. *Renew. Sustain. Energy Rev.* **2018**, *82*, 3323–3332. [CrossRef]

7. Verzijlbergh, R.A.; De Vries, L.J.; Lukszo, Z. Renewable Energy Sources and Responsive Demand. Do We Need Congestion Management in the Distribution Grid? *IEEE Trans. Power Syst.* **2014**, *29*, 2119–2128. [CrossRef]

8. Clement-Nyns, K.; Haesen, E.; Driesen, J. The Impact of Charging Plug-In Hybrid Electric Vehicles on a Residential Distribution Grid. *IEEE Trans. Power Syst.* **2010**, *25*, 371–380. [CrossRef]

9. PricewaterhouseCoopers. *Smart Charging of Electric Vehicles: Institutional Bottlenecks and Possible Solutions*; Technical Report; PricewaterhouseCoopers: Utrecht, The Netherlands, 2017.

10. Sortomme, E.; El-Sharkawi, M.A. Optimal Scheduling of Vehicle-to-Grid Energy and Ancillary Services. *IEEE Trans. Smart Grid* **2012**, *3*, 351–359. [CrossRef]

11. van der Kam, M.; van Sark, W. Smart charging of electric vehicles with photovoltaic power and vehicle-to-grid technology in a microgrid: A case study. *Appl. Energy* **2015**, *152*, 20–30. [CrossRef]

12. Sarantis, I.; Alavi, F.; Schutter, B.D. Optimal power scheduling of fuel-cell-car-based microgrids. In Proceedings of the 2017 IEEE 56th Annual Conference on Decision and Control (CDC), Melbourne, Australia, 12–15 December 2017; pp. 5062–5067. [CrossRef]

13. Knap, W. *Horizon at Station Cabauw*; Pangaea: Cabauw, The Netherlands, 2007. [CrossRef]

14. Stein, J.S.; Holmgren, W.F.; Forbess, J.; Hansen, C.W. PVLIB: Open source photovoltaic performance modeling functions for Matlab and Python. In Proceedings of the 2016 IEEE 43rd Photovoltaic Specialists Conference (PVSC), Portland, OR, USA, 5–10 June 2016; pp. 3425–3430. [CrossRef]

15. Apostolaki-Iosifidou, E.; Codani, P.; Kempton, W. Measurement of power loss during electric vehicle charging and discharging. *Energy* **2017**, *127*, 730–742. [CrossRef]

16. Shareef, H.; Islam, M.M.; Mohamed, A. A review of the stage-of-the-art charging technologies, placement methodologies, and impacts of electric vehicles. *Renew. Sustain. Energy Rev.* **2016**, *64*, 403–420. [CrossRef]

17. Daina, N.; Sivakumar, A.; Polak, J.W. Modelling electric vehicles use: A survey on the methods. *Renew. Sustain. Energy Rev.* **2017**, *68*, 447–460. [CrossRef]

18. Francfort, J. *EV Project Data & Analytic Results*; Idaho National Laboratory: Idaho Falls, ID, USA, 2014.

19. Bemporad, A.; Morari, M. Robust model predictive control: A survey. In *Robustness in Identification and Control*; Garulli, A., Tesi, A., Eds.; Lecture Notes in Control and Information Sciences; Springer: London, UK, 1999; pp. 207–226.

20. Bemporad, A.; Morari, M. Control of systems integrating logic, dynamics, and constraints. *Automatica* **1999**, *35*, 407–427. [CrossRef]

21. IEEE. *C57.91-2011—IEEE Guide for Loading Mineral-Oil-Immersed Transformers and Step-Voltage Regulators*; IEEE: Piscataway, NJ, USA, 2012; pp. 1–123. Available online https://ieeexplore.ieee.org/document/6166928 (accessed on 18 November 2019). [CrossRef]

22. Smith, M.; Castellano, J. *Costs Associated with Non-Residential Electric Vehicle Supply Equipment: Factors to Consider in the Implementation of Electric Vehicle Charging Stations*; Technical Report DOE/EE-1289; U.S. Department of Energy Vehicle Technologies Office: Washington, DC, USA, 2015.

23. Blok, R.F.P. *Verslag Benchmark Publiek Laden 2018—Sneller naar een Volwassen Markt*; Technical Report; NKL Nederland: Emmen, NL, USA, 2018.

Review

Integration of Electric Vehicles in the Distribution Network: A Review of PV Based Electric Vehicle Modelling

Asaad Mohammad, Ramon Zamora and Tek Tjing Lie *

Department of Electrical and Electronic Engineering, Auckland University of Technology, Auckland 1010, New Zealand; asaad.mohammad@aut.ac.nz (A.M.); ramon.zamora@aut.ac.nz (R.Z.)
* Correspondence: tek.lie@aut.ac.nz

Received: 23 July 2020; Accepted: 31 August 2020; Published: 2 September 2020

Abstract: Electric vehicles (EVs) are one of a prominent solution for the sustainability issues needing dire attention like global warming, depleting fossil fuel reserves, and greenhouse gas (GHG) emissions. Conversely, EVs are shown to emit higher emissions (measured from source to tailpipe) for the fossil fuel-based countries, which necessitates renewable energy sources (RES) for maximizing EV benefits. EVs can also act as a storage system, to mitigate the challenges associated with RES and to provide the grid with ancillary services, such as voltage regulation, frequency regulation, spinning reserve, etc. For extracting maximum benefits from EVs and minimizing the associated impact on the distribution network, modelling optimal integration of EVs in the network is required. This paper focuses on reviewing the state-of-the-art literature on the modelling of grid-connected EV-PV (photovoltaics) system. Further, the paper evaluates the uncertainty modelling methods associated with various parameters related to the grid-connected EV-PV system. Finally, the review is concluded with a summary of potential research directions in this area. The paper presents an evaluation of different modelling components of grid-connected EV-PV system to facilitate readers in modelling such system for researching EV-PV integration in the distribution network.

Keywords: plug-in electric vehicle; energy management system; renewable energy; vehicle-to-grid

---

1. Introduction

The issues like global warming, depleting fossil fuel reserves, and greenhouse gas (GHG) emissions need dire attention for ensuring a sustainable future. Because the transportation sector is one of the largest contributors to the rising harmful emissions, the electrification of transportation is seen as a promising solution for this problem. Electric vehicle (EV) technology has existed for more than a century peaking commercially around 1900. However, due to the easy availability of fossil fuels, advancements in internal combustion (IC) technology, and simplicity in the use of IC engines, EVs were put on hold and limited to golf carts and delivery vehicles. Figure 1 shows the progression timeline of the EVs. The dependency on petroleum imports for transportation purposes is also reduced by electrification of transportation, thereby increasing energy security. However, the adoption rate of EVs remains slow owing to factors, such as high initial cost, battery degradation, inadequate charging infrastructure, range anxiety, etc. [1]. Various policies and incentives are made available by governments around the world to promote the uptake of EV and to prevent these barriers from realizing a complete shift to electrified transportation. As per the report "Global EV outlook" of the International Energy Agency, the total number of EVs are projected to reach 130 million by 2030 [2].

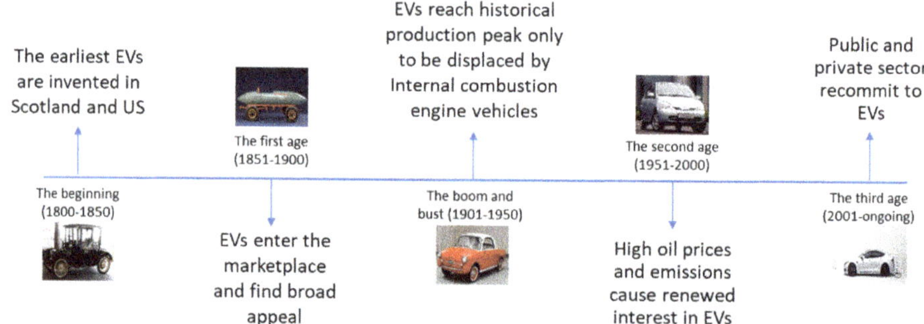

Figure 1. The evolution of electric vehicles (EVs).

However, high penetration of EVs also poses distribution network quality issues, particularly network congestion, three-phase voltage imbalance and off-nominal frequency problems. The EVs are a mobile single-phase load so they can be randomly plugged in at any one of three phases within distribution networks, leading to a scenario that electrical components in one particular phase, such as power supply cable, overhead line or transformer may be heavily loaded while the rest of two phases are not. The unbalanced three-phase loading may lead to a series of negative impact on power quality issue: Transformer failures, equipment loss-of-life, relay misfunction, etc. Moreover, as EVs are highly spatial and temporally uncertain, handling EVs as additional loads while maintaining the reliability and security of the grid is difficult. The coincidence of timing between EV home charging and residential load peaks leads to additional system peaks. Moreover, multiple EV chargers in a neighbourhood can introduce significant harmonics, thereby reducing power quality [3]. Therefore, the integration of substantial EV penetration in the distribution networks is a significant area of interest in the research and engineering community, especially optimally controlling EV charging to minimise the impact of the above-described issues.

Another significant contributor to harmful emissions is the power industry, particularly fossil fuel-based power generation. Renewable energy sources (RES), such as wind and solar are increasingly adopted to mitigate the power industry emissions. The variable nature of RES which depends on the weather, time, location, etc. creates voltage stability and reliability issues for the power grid requiring integration of Energy Storage System (ESS). Also, there may not be sufficient demand requirement during the period of high RES generation, which leads to the under-utilisation of average generated capacity. Using ESS with RES can result in its effective utilisation as ESS can store energy when demand is low and supply back when demand is high. Apart from using ESS, application of demand-side management techniques like load shifting, time of use pricing, and demand bidding can also solve the aforementioned problems associated with RES although the impact of these techniques is limited compared to ESS [4,5].

Large-scale integration of RES requires an increased size (or capacity) of ESS. Hence, it leads to a significant capital requirement, especially due to the high per-unit cost of ESS. As we are already moving towards electric vehicles to combat GHG emissions and these EVs essentially run on the batteries, the EVs can also act as a dynamic natured ESS, due to the vehicle-to-grid (V2G) feature, in which EVs deliver energy stored in their batteries back to the grid [6]. Additionally, EVs spend a considerable amount of time (22 h) in parked conditions [7], so they can be suitably used as ESS without creating inconvenience (e.g., range anxiety issues) for users. However, battery degradation is still an issue which can be offset by giving incentives to users/aggregators to participate in V2G. As the battery capacity of each EV is minuscule compared to grid load requirements, an aggregation of EVs is generally required to provide the grid with the backup power. Apart from storing surplus energy generated by RES, EVs can also provide the grid with additional ancillary services, such as voltage regulation, frequency regulation, spinning reserve, etc. EVs can also participate in energy trading, to be a source of revenue for the aggregator/users to compensate for the battery degradation, due to participation in V2G. However, most of the energy markets around the world require a minimum capacity to participate, which would require an aggregator of a large number of EVs. To counteract this, more research is being done on transactive or peer-to-peer (P2P) trading mechanisms [8].

Moreover, the emission benefits of electrified transportation cannot be maximised if the source of EV charging is based on non-renewable sources. In fact, EVs are shown to emit higher emissions, measured from well to wheel, i.e., source to the tailpipe for the countries whose primary source of power generation is based on fossil fuels [9]. However, using RES to charge the EVs could result in reducing GHG emissions, as shown in Reference [10], where 50,000 EVs charged from a mix of wind and PV energy sources resulted in 400 Mtons less emissions per year.

Based on these factors, this paper presents a general framework for designing a grid-connected EV-PV system. Several papers have also reviewed the different aspects of the interaction of EV-PV system and distribution network in the literature. References [11–14] discuss charging EVs using PV generation with a focus on control architectures and algorithms, and economic framework. The impact of the charging infrastructure of EV on the grid in terms of power quality is reviewed in Reference [15]. An overview of EV modelling techniques is presented in References [16–18] with an emphasis on modelling methods for EV loads and charging stations.

These review papers study the limited aspects of the interaction of grid-connected EV with RES, particularly PV, focusing on the modelling of control methods or EV loads. Also, a detailed review of modelling the uncertainties present in the grid-connected EV-PV system is not present in the literature to the knowledge of the authors. Therefore, this paper presents a comprehensive review of all aspects of modelling a grid-connected EV-PV system viz., control architectures, charging algorithms, and uncertainty analysis. This paper aims to provide an evaluation of these aspects to enable the researchers to model a grid-connected EV-PV system for carrying out impact or implementation studies of EV integration into the distribution system. The grid is represented by a distribution network as EV and PV both are on the distribution side. Throughout the paper, EV-PV system is considered as a single entity (limited to the times when connected to the grid for charging or vehicle-to-grid), and the PV is considered as a complementary energy source to charge EVs other than the grid. Figure 2 shows the analytical framework of the modelling aspects of grid-connected EV-PV system.

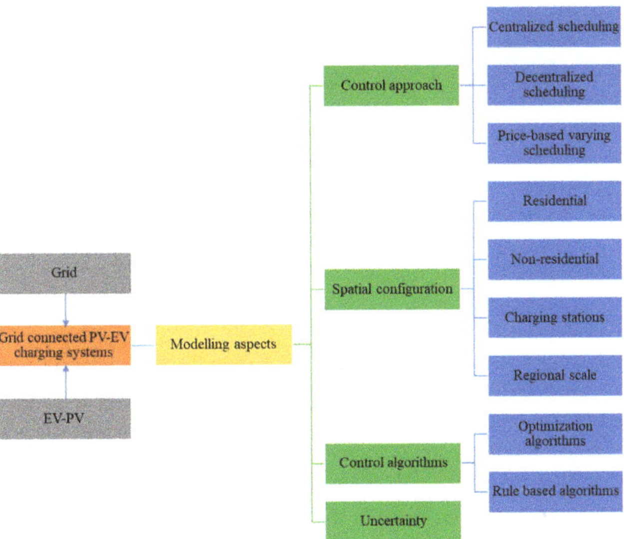

Figure 2. An analytical framework for grid-connected EV-PV (photovoltaics) interaction.

The organisation of the paper is as follows: Section 2 provides an overview of the modes of EV integration with the grid. Section 3 discusses the control architectures of connecting EVs to the grid. Section 4 describes the state-of-the-art literature of smart charging algorithms of grid-connected EV-PV system. Section 5 reviews the uncertainty analysis methods for EV demand, PV generation, and load distribution. The suggestions for future research with concluding remarks are presented in Section 6.

2. EV Interaction with the Distribution Network

Figure 3 shows a general representation of an EV connected to the electrical grid. The technology which allows the bidirectional flow of energy between EV and grid is known as vehicle-to-grid (V2G). It is achieved by the integration of Information and Communication Technologies (ICT) with the EV charging system. The modelling research of EV interaction with the distribution network has transitioned from unidirectional mode in the initial stage to bidirectional mode in the current stage [6,7].

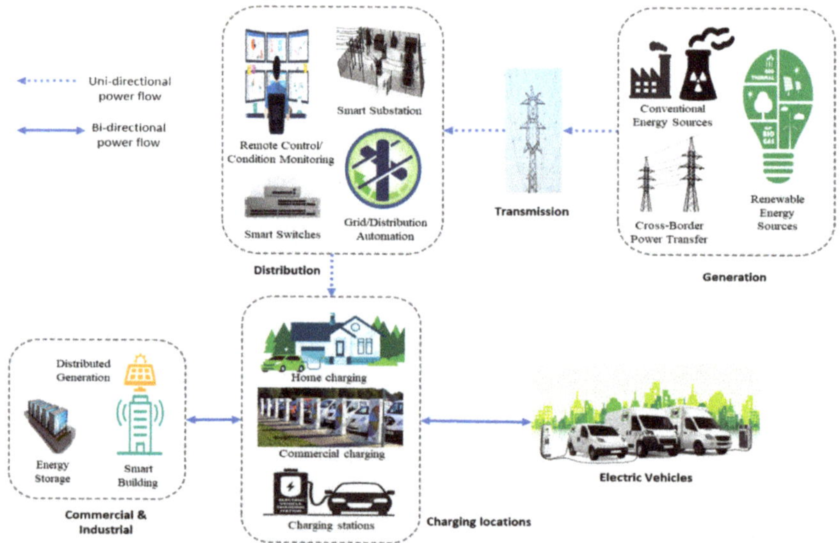

Figure 3. EV integration with the electrical grid.

With the increasing level of EV penetration, the associated technical issues, e.g., system imbalance, decreased stability, and power quality, as well as increased system cost, are becoming more prominent, due to additional energy and power demand. The unidirectional approach, i.e., G2V mode, has been extensively studied in the literature in the form of topics like smart charging [19], safety [20], and control features [21]. The focus of these studies is on minimizing the charging cost [22] or minimizing the impact on the distribution system [23,24].

However, in the bidirectional mode, EV is not only the load for the grid, but also a distributed generation and storage. The initial idea was to use EV battery to store energy and send it back to the grid in peak period, known as peak load shaving [6]. Reference [25] presents a review of peak shaving strategies using demand-side management, energy storage systems, and electric vehicles. Table 1 illustrates the characteristical differences between the unidirectional and bidirectional modes. As an individual EV has a small battery capacity, a major challenge is the synchronisation of a large number of EVs charging/discharging operation required for them to be an effective storage system. Also, the limited uptake of EV did not quite make this idea of using EV in the bidirectional mode mainstream. Research later indicated that the application of bidirectional V2G in the ancillary market: Spinning reserve and voltage control is much more important than peak load reduction. Spinning reserve is the extra generation that can be made readily available, and it is paid for the availability along with the time it is called for deployment (compared to peak load shaving), which makes deployment of EV in ancillary service provision very economically favourable. Moreover, in terms of frequency of deployment, the voltage regulation is needed more than 300 times per day compared to the need for peak load shaving, which is only a few hundred hours per year [26].

Table 1. Modes of Interaction between EV and grid.

Features	Unidirectional	Bidirectional
Power flow	Grid-to-vehicle (G2V)	G2V and vehicle-to-grid (V2G)
Infrastructure	Communication	Communication, bidirectional charger
Cost	Low	High
Complexity	Low	High
Services	Load profile management, Frequency regulation [27]	Backup power support, frequency regulation, voltage regulation, active power support [28]
Advantages	Overloading prevention, load levelling, profit maximisation, emission minimisation [29]	Overloading prevention, profit maximisation, emission minimisation, renewable energy sources (RES) integration, voltage profile improvement, harmonic filtering [30], load levelling, power loss reduction [31]
Disadvantages	Limited services	Battery degradation, high complexity, and cost, social barriers

Initially, V2G involved only energy transfer from EVs to the distribution system. However, with the advancement in technology, two new energy transfer modes (V2H and V2V) are added. Therefore, the bi-directional energy transfer from EV can now be classified into:

- Vehicle-to-grid (V2G): Energy transfer from EV to the distribution network.
- Vehicle-to-home/building (V2H/V2B): Energy transfer from EV to home/building.
- Vehicle-to-vehicle (V2V): Energy transfer from one EV to another EV.

3. Modelling of Grid-Connected EV-PV System

The sustainability of EV depends on the source of charging. All forms of EVs, i.e., plug-in electric vehicle (PEV), hybrid electric vehicle (HEV), or plug-in hybrid electric vehicle (PHEV), have lower emissions if the energy supplied for charging is based on clean fuel, such as renewable sources. However, contrary to popular belief if the EVs are charged from fossil fuel or gas-based generation, the emissions are significant and not zero. The RES, i.e., PV, wind, tidal, geothermal, or hydro, are excellent options to power electric vehicles. Moreover, the following reasons make PV an admirable source to charge the EVs:

- The cost of PV has been dropping continuously and is currently less than $1/W_p$ [32].
- PV is highly accessible, i.e., PV modules are generally installed on the building rooftops and carparks, close to EV locations.
- PV modules do not require maintenance and are also noise-free.
- EVs can store the surplus generated solar energy, thereby eliminating the need for battery systems [33,34].

Figure 4 shows a general framework for designing a smart charging system for integrating EV-PV system into the grid. As the focus of this review paper is on modelling aspect of the grid-connected EV-PV system, this section will provide an overview of the modelling of control approaches with subsequent sections reviewing about charging models/algorithms and uncertainty.

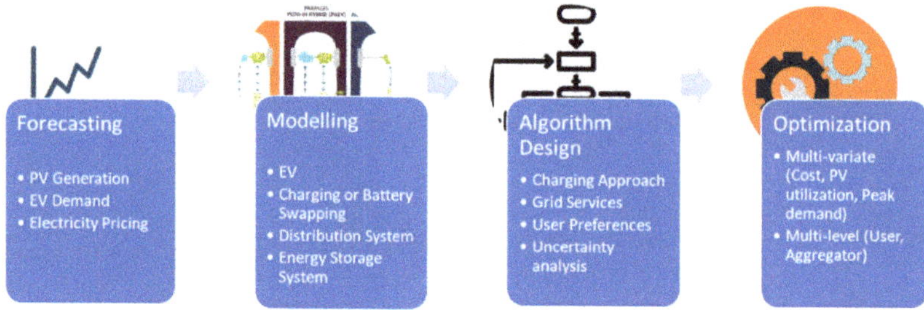

Figure 4. A general outline for modelling a grid-connected EV-PV charging system.

The control architectures for grid-connected EVs (with or without PVs) can be categorised into the following three methodologies:

- Centralised scheduling;
- Decentralised scheduling;
- Price-varying scheduling.

In centralised scheduling method, EV aggregator plays a crucial role in integrating EV with the grid. Initially, each EV sends the necessary charging related information to the aggregator. After which aggregator computes the optimal charging strategy and participates in the energy trading through bidding, which is verified by grid system operators. The general objective functions in centralised type scheduling are charging cost minimisation [35], line power loss minimisation [36], aggregator profit maximisation [37,38], voltage regulation [39] and frequency regulation [40]. Due to the aggregation of many EVs, this method is very good for providing backup power and ancillary services. However, in the centralised method, EV users have to relinquish the charging process control to centralised authority. Other drawbacks of this approach are high dependency on the control centre and large communication bandwidth.

In decentralised scheduling method, individual EVs are controlled directly instead of through a central control unit. Firstly, EV aggregator formulates a bidding strategy based on EV load demand data collected or forecasted in a given period. Then, the bids are submitted to the central grid operator and cleared in the energy market the same as in centralised scheduling. After the bids are approved, and an agreement is done with the grid operator, the aggregator broadcasts the charging prices to individual EV users. Based on the price and convenience, users decide whether to charge/discharge their EVs in a given period. The advantage of this type of scheduling is that the infrastructure is simple and of low cost. However, due to a random number of EVs guaranteed to be available at a given time, this method's capability of the provision of backup power and ancillary services is low. Also, privacy and security issues are there. The general objective function in decentralised type scheduling is mainly charging cost minimisation [41–43]. Other objectives are RES integration [44], load profile levelling [45], voltage regulation [46] and frequency regulation [47].

The price-varying scheduling has the same structure as decentralised scheduling, however, the charging behaviours of EVs are directly affected by varying electricity pricing. Instead of two-way communication, i.e., price and power schedule information exchanged in decentralised scheduling, here the only price is communicated to EVs. Reference [48] discusses the feasibility of using time-of-use (TOU) based pricing for EV energy management. Reference [49] presents a socially optimal pricing system between EV aggregators and utility. Reference [50] introduces a smart EV energy management algorithm that takes dynamic factors, such as user participation and load variation into account.

Figure 5 presents an overview of the comparison between the scheduling strategies discussed above [51]. Even though the price-varying scheduling is overall less complex, it is less attractive for commercial entities to participate in V2G, due to the high cost of computation on their side. Hence, the focus of research is generally more on centralised and decentralised scheduling strategies.

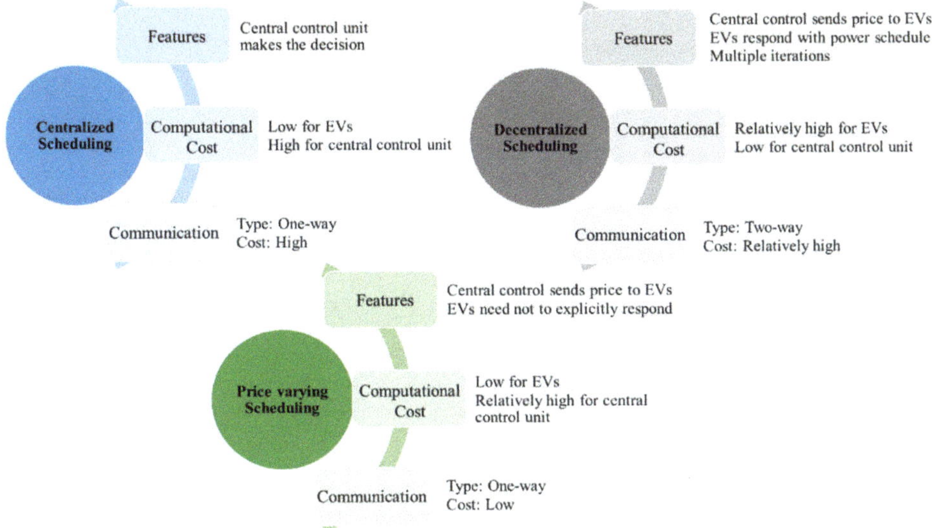

Figure 5. An overview of scheduling strategies in V2G mode [51].

The grid-connected EV-PV systems are designed based on spatial configuration requirements, i.e., for homes or office use etc. Generally, in the literature, four space-based levels are used: Residential (individual house), non-residential (commercial/workplaces), public charging stations and inter-territory region. Due to the large size of EV loads, which almost doubles the electricity consumption of a household, it is reasonable to provide another energy source (like PV) [52]. Nevertheless, it appears through the literature that while coupling EV with PV inside households can be beneficial, the benefits are bounded by the EV utilisation for mobility. Most of the EVs are usually away from home during the day, and therefore, cannot benefit from maximal PV generation. It is reasonable to assume that usually, EVs will be at non-residential places (commercial/workplaces) during this day period when peak PV generation happens. So, EVs will be either at residential or non-residential areas. Therefore, the focus of this paper is only on the modelling of residential and non-residential (commercial/workplace) EV-PV system. The PV based EV charging stations are not yet economically feasible, due to the marginal cost associated with PV generation and the cost of energy storage systems [53]. Reference [54] is one example of the limited literature available on standalone PV based EV charging stations.

4. EV Smart Charging Using PV and Grid

Multiple studies have explored the advantages of a PV based EV charging system. Reference [55] demonstrates the advantage of using PV to charge the EV and show that it allows for greater penetration of both PV and EV. EVs can also mitigate the negative effects of excess PV generation [56]. Reference [57] presents a case study of Columbus, USA, in which it is demonstrated that charging EV from the PV is more economical and produces less CO_2 footprint than charging EV from the grid. A case study presented in Reference [58] compares charging of EVs through the modes: Only grid, only PV with battery storage and grid integrated PV and finds that the grid integrated PV performs better economically compared to the other two systems. In Reference [59], the authors discuss the application of PV energy and EV as an energy storage system to mitigate the peak loading in the grid. These studies demonstrate the advantages of PV based EV charging over grid EV charging. There is a vast amount of literature on different charging algorithms or achieving different economic, technical, or social objectives related to PV based EV charging. Table 2 provides a summary of key smart charging related works for the grid-connected EV-PV system. The optimisation model type depends on the problem formulation. Generally, the convex type problems (linear, mixed-integer, quadratic) can achieve optimal solutions with a low computational cost. For non-convex problems, meta-heuristic type optimisation methods (Genetic Algorithm, Particle Swarm Optimisation) are useful to achieve a near-optimal solution with a low computational burden. The rule-based algorithm or heuristic type optimisation methods can produce good enough solutions for random instantaneous events (e.g., plugging/unplugging of EVs, PV power variation) with little data and computational power requirements. The focus of the literature is generally on residential or office PV based EV charging system, not on commercial applications, due to less complexity in analysis and modular integration in the distribution system. Moreover, almost all the smart charging research focuses on the specific aspects of optimizing the EV integration into grid, e.g., slow/fast charging, market participation, ancillary services. For emulating the real-life implementation, a comprehensive system with multiple aspects is required. Reference [60,61] are some early stages work on a comprehensive system combining multiple aspects which are usually studied in isolation.

Table 2. Summary of literature related to smart charging of grid-connected EV-PV system.

References	Objectives	Optimisation Model	Software/Implementation	Key Findings
[62]	Peak shaving and valley filling	Linear programming	MATLAB	The effectiveness of the proposed algorithm is dependent on a high number of available parking spots.
[35]	Maximizing profit and PV utilisation	Mixed Integer Linear programming	GAMS	Due to battery degradation cost, V2G is not economically feasible unless high PV production is present
[63]	Minimizing system cost	Mixed Integer Linear programming	CPLEX	Smart charging can result in saving of operational cost for charging and PV usage for the parking lot owner
[64]	Minimizing charging cost	Fuzzy logic	MATLAB	The algorithm is not optimisation based so targets several objectives: Reduction in charging cost and system losses, improvement in voltage profile.

Table 2. *Cont.*

References	Objectives	Optimisation Model	Software/Implementation	Key Findings
[65]	Maximizing PV utilisation	Metaheuristic	MATLAB	The proposed heuristic algorithm achieves desired objectives with low computational cost and without forecasting of uncertain variables.
[66]	Maximizing EV aggregator benefits	Hybrid MPC	-	The proposed algorithm achieves near-optimal solution of EV charge scheduling problem with better efficiency than standard MPC
[67]	Maximizing PV utilisation and reducing EV charging impact	Linear programming	Case study: New South Wales distribution system	The proposed strategy controls the charging/discharging profile of EVs to match with the shape of the PV output to achieve desired objectives.
[68]	Minimizing charging cost	Mixed Integer Linear programming	Case study: Korea	The proposed algorithm does not consider selling excess power and demonstrates charging cost savings compared to uncoordinated charging
[61]	Minimizing system cost	Mixed Integer Linear programming	Microsoft Solver Foundation	A comprehensive system to achieve one optimal charging profile will result in a larger net benefit compared to individual applications.
[69]	Minimizing charging cost	Convex programming	MATLAB	ESS can significantly reduce charging cost and bi-directional V2H is cheaper than H2V
[70]	Maximizing profit and ESS life	Non-linear programming	GAMS	Considering only revenue maximisation will result in an adverse effect on ESS life
[71]	Maximizing PV utilisation	Linear programming	Case study: LomboXnet	Proposed algorithm increases PV self-consumption and reduces peak demand by half
[72]	Minimizing charging cost	Rule-based algorithm	MATLAB	Rule-based charging is superior to conventional charging for less charging cost and reduced grid loading
[73]	Maximizing PV utilisation	Rule-based algorithm	MATLAB	V2B can be an effective strategy if initial capital costs and electricity price are fitting
[74]	Minimizing peak demand	MPC	MATLAB	EV scheduling can reduce both the magnitude and frequency of peak loading
[75]	Peak shaving and valley filling	Quadratic programming	MATLAB	Net load variation was lower in case of low PV power-sharing and vice-versa

V2H, vehicle-to-home; H2V, home-to-vehicle; V2B, vehicle-to-building; MPC, model predictive control.

The stochastic behaviour of the PV generation is a major disadvantage for EV charging. The approach of a smart charging algorithm is to provide flexibility in EV charging to account for the uncertainty in PV generation. Reference [71] has shown that smart charging, along with the V2G technology, increases PV self-consumption and reduces peak demand. Reference [76] varies the EV charging power with time to match with the generated PV power and achieves the condition of maximum PV utilisation. Another way to counteract uncertainty is the sequential charging in which the total number of EVs charging at constant power is varied dynamically so that the net charging power follows the PV generation, as seen in Reference [77]. Reference [78] considers multiple cases to show the superiority of sequential charging over concurrent charging in terms of PV utilisation under stochastic conditions. However, due to no associated time constraints, it is not feasible for workplace charging.

5. Uncertainty Modelling

This section reviews the methods for modelling the uncertainties present with the various input parameters for the EV-PV grid integrated system. Three input factors are of main interest: EV charging demand, PV generation, and Electrical load distribution. The tables in respective sections summarise the techniques used to model the uncertainties present. The remarks show the comparative analyses of these techniques in terms of system size, computational cost, and accuracy.

5.1. EV Charging Demand

The uncertainties in EV charging demand are due to multiple factors, e.g., user behaviour, charging infrastructure, and operational parameters. Table 3 presents an overview of various uncertainty methods for modelling EV load demand in terms of application and associated drawbacks. Generally, Monte Carlo and Probability distribution based modelling method is common practice in the literature. However, due to computational cost and accuracy issues associated with them, respectively, more advanced methods like Markov chain and Information gap decision theory are used for specific applications. A hybrid approach of combining methods is also used to minimise the associated drawbacks.

Table 3. Overview of uncertainty modelling methods for EV load demand.

Method	Remarks	References
Scenario reduction	• Simple and less computationally intensive • Approximate uncertainty modelling, accuracy depends on the amount of historical data available	[79,80]
Monte Carlo simulation	• High accuracy, but also computationally intensive • Accuracy depends on the amount of historical data available	[19,81]
Fuzzy logic	• Historical data not required • Accuracy depends on rule settings which are based on researcher experience	[82,83]
Hybrid Monte Carlo-fuzzy	• High accuracy, but also computationally intensive • Can model both temporal and spatial uncertainty	[84]
Artificial Neural Network	• Accuracy depends on input dataset quality • Considers the correlation between forecasted and observed data	[37,85]
Markov chain	• Very high accuracy with moderate computational cost • Performance depends on input data dimension	[86]
Probability distribution fitting	• Very simple, but also less accurate	[87,88]
Robust optimisation	• Low computationally intensive however difficult to employ with non-linear problems • Not flexible, i.e., give a single solution which might be infeasible	[89,90]
Information gap decision theory	• Useful for dealing with severe uncertainties • Complex implementation	[91,92]

Figure 6 shows the various input parameters for the uncertainty modelling of EV load demand. The parameters related to time (e.g., arrival, departure, travel, service) and charging power demand required are common in all the three modes of charging: Individual, residential, and commercial, while others are specific to the application. The uncertainties in the parameters involving human factors, i.e., travel/arrival/departure time and pattern are difficult to describe accurately, and also the literature is quite scarce on the effect of human learning capability on EV charging demand. Reference [44] is an example of paucity of research on the practical effect of human factors on EV charging.

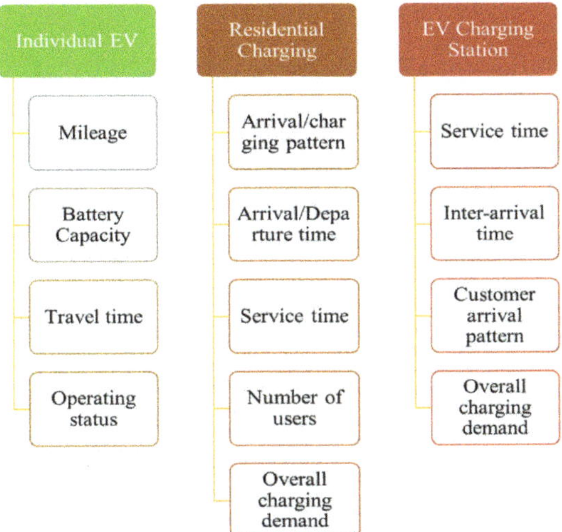

Figure 6. EV load demand parameters used for uncertainty modelling.

5.2. PV Generation

A PV module converts energy from the sun into electrical form depending upon the incident radiation on the module surface. This incident solar radiation is highly variable and depends on various geographical and metrological factors. The common variables used in uncertainty modelling of PV generation are solar irradiance, sky type index (clear, cloudy, sunny), module and air temperature, wind speed, and humidity. Table 4 shows a summary of commonly used uncertainty modelling methods for PV generation. The commonly practiced methods are Point estimation, Monte Carlo, Scenario based analysis, and statistical methods (Autoregressive Moving Average). These methods are less complex and work well with small system size. However, for bigger PV systems, Rolling Horizon approach and Kernal Density are more suitable. Generative Adversarial Network (GAN) is the latest uncertainty modelling method based on a machine learning approach.

Table 4. Overview of common uncertainty modelling methods for PV.

Method	Remarks	References
Point estimation	• Computationally intensive with more input variables	[93]
Bootstrap	• Simple and low computational cost • High accuracy	[94]
Monte Carlo simulation	• High accuracy, but also computationally intensive	[95]
Mean-Variance estimation	• Based on the assumption that uncertainty is normally distributed • Simple, but less accurate for practical cases	[96]
Two stage scheduling	• Upper level deals with global adjustment and lower with local adjustment • Simple, flexible and accurate	[97]
Scenario based analysis	• Very commonly used method with a high degree of accuracy • Accuracy depends upon the scenario generation technique	[98]
Kernel Density estimation	• Needs to analyse a large amount of historical data	[99]
Autoregressive Moving Average	• Accuracy depends on historical time-series dataset • Needs a lot of historical data and analysis	[35]
Probability distribution fitting	• Very simple, but also less accurate	[100]
Rolling Horizon approach	• Effective for large scale system with moderate computational cost	[101]
Generative Adversarial network	• Very new and highly accurate scenario based method	[102]

References [103,104] describes the implementation details of various forecasting techniques for PV power generation. More details about uncertainty modelling for the RES systems can be found in References [105–108]. The literature of PV based uncertainty modelling is scarce as the cumulative effect of PV power on the system is small compared to other uncertain variables (load, EV demand).

The most common method to mitigate the PV uncertainty is using an external battery storage system, i.e., different from the EV batteries [109]. The excess PV generation, usually in the afternoon, is stored in the battery pack and used to charge the EVs when PV generation is inadequate. Reference [110] compares three different algorithms for finding the best operation characteristics for the battery storage and finds that using a sigmoid function-based discharging algorithm, while charging EVs during the night and storing PV excess is the best approach. However, these studies do not consider the optimal sizing of the external battery storage system as it is a quite expensive component. Apart from mitigating PV uncertainty, the external battery storage system also minimises the impact of EV demand uncertainty parameters constrained by time.

5.3. Electrical Load Demand

The consumption of electricity is highly spatially and temporally uncertain, varying between different load sources, seasons, and the time of day. The main factors for introducing uncertainty in load sources are user behaviours, climatic conditions, and electrical equipment variation [111]. Table 5 shows an overview of various common methods used for modelling uncertainty in electrical loads. Readers can refer to [105,108,111–113] for implementation details of these and other methods used to model uncertainties present in electrical load. The convolution and cumulant based techniques are traditional methods popular in the late nineties' era. However, with the scaling of computational cost with system size, the point estimation became a more popular method. Monte Carlo and Scenario based analysis are also fairly common in the literature.

Table 5. Overview of common uncertainty modelling methods for electrical load demand.

Method	Remarks	References
Point estimation	• Does not require complete knowledge about the system, but computationally intensive with more input variables • Two-point is the simplest and three-point is most efficient	[114,115]
Monte Carlo simulation	• High accuracy, but also computationally intensive • Different sampling techniques reduce the computational burden	[116,117]
Fuzzy logic	• Less computationally intensive and robust in nature • Vital parameters are decided by the researcher based on experience	[118]
Scenario based analysis	• Very commonly used method with a high degree of accuracy • Accuracy depends upon the scenario generation technique	[119]
Autoregressive Moving Average	• Accuracy depends on historical time-series dataset • Needs a lot of historical data and analysis	[85,120]
Convolution based	• Traditional analytical method with low computation efficiency • Applicable to linear systems with independent inputs	[121,122]
Probability distribution fitting	• Very simple, but also less accurate	[88,123]
Cumulant based	• Traditional analytical method with high computation efficiency • Accuracy decreases with higher order systems	[124]

6. Conclusions and Future Research Suggestions

Electric vehicles and renewable energy-based generation are a promising solution to rising GHG emissions. Further, EVs can act as a dynamic energy storage system through the technology of V2G, thereby, facilitating RES integration in the smart grid. Also, well to wheel emissions from EVs depend upon the charging source. Therefore, RES based EV charging is desired for the overall reduction in emissions and getting the best of both technologies. Thus, this research area is quite popular and needs further exploration for worldwide implementation. This paper presents a state-of-the-art comprehensive review of the modelling of grid-connected EV-PV charging systems. A general framework of designing the grid-connected EV-PV system is described along with a focus on smart charging algorithms. The modelling techniques for associated uncertainties with the grid-connected EV-PV system, i.e., EV demand, electrical load, and PV generation are also intensely reviewed. The study reveals that although the research in this area is plentiful, few gaps need to be investigated. Some future research directions are suggested as following:

- Smart charging algorithms

The EV charging models need to be more comprehensive in nature, i.e., multiple charging powers, charging station and battery-swapping station, and wholesale market trading and ancillary services provisions, in order to more accurately and realistically model the practical implementation. More studies with respect to finding the optimal trade-offs between computational burden and performance should be made.

- P2P V2G power transfer

There is a need for more research on peer-to-peer or transactive type charging systems as this encourages all types (big, small, etc.) EV aggregators to trade energy with one another instead of only sizeable aggregator participating in central energy trading. Another advantage is that transactive trading can operate independently of direct influence from the grid so that the price signal from the central power station may not affect the performance of the transactive trading the way it influences the scheduling and trading of energy in existing systems.

- Uncertainty analysis

The focus of future research should be on finding more realistic forecasting and uncertainty analysis techniques that optimally balance simplicity and performance. Also, more advancement is needed in the modelling of challenging variables like human behaviour, etc. Further, almost all the current research focuses on improving PV forecasting accuracy rather than addressing uncertainties associated with PV generation.

- PV based EV charging stations

With PV based EV charging being a viable solution for emission issues, more research is needed on the commercial aspects, e.g., solar charging stations as current research focus more on residential EV-PV systems. The impact analysis and interaction with the distribution system needs to be studied in detail.

- Price-varying scheduling

Because of easy implementation and effectiveness for managing charging load in peak/valley times, price-varying scheduling is very attractive to aggregators. Therefore, more research is required for charging models based on price response and price elasticity.

Author Contributions: Conceptualization, A.M. and R.Z.; methodology, software, validation, formal analysis, investigation, resources, data curation, writing—original draft preparation, A.M.; writing—review and editing, visualization A.M., R.Z., and T.T.L.; supervision, project administration, R.Z. and T.T.L. All authors have read and agreed to the published version of the manuscript.

Funding: This research received no external funding.

Conflicts of Interest: The authors declare no conflict of interest.

References

1. Asaad, M.; Shrivastava, P.; Alam, M.S.; Rafat, Y.; Pillai, R.K. Viability of xEVs in India: A public opinion survey. In *Lecture Notes in Electrical Engineering*; Springer: Singapore, 2018; Volume 487, pp. 165–178. ISBN 9789811082481.
2. Bunsen, T.; Cazzola, P.; Gorner, M.; Paoli, L.; Scheffer, S.; Schuitmaker, R.; Tattini, J.; Teter, J. *Global EV Outlook 2018: Towards Cross-Modal Electrification*; International Energy Agency: Paris, France, 2018.
3. Monteiro, V.; Gonçalves, H.; Afonso, J.L. Impact of Electric Vehicles on power quality in a Smart Grid context. In Proceedings of the 11th International Conference on Electrical Power Quality and Utilisation, Lisbon, Portugal, 17–19 October 2011; pp. 1–6.
4. Jordehi, A.R. Optimisation of demand response in electric power systems, a review. *Renew. Sustain. Energy Rev.* **2019**, *103*, 308–319. [CrossRef]
5. Strbac, G. Demand side management: Benefits and challenges. *Energy Policy* **2008**, *36*, 4419–4426. [CrossRef]
6. Kempton, W.; Letendre, S.E. Electric vehicles as a new power source for electric utilities. *Transp. Res. Part D Transp. Environ.* **1997**, *2*, 157–175. [CrossRef]
7. Paşaoğlu, G.; Fiorello, D.; Martino, A.; Zani, L.; Zubaryeva, A.; Thiel, C. Travel patterns and the potential use of electric cars—Results from a direct survey in six European countries. *Technol. Forecast. Soc. Chang.* **2014**, *87*, 51–59. [CrossRef]

8. Tushar, W.; Yuen, C.; Mohsenian-Rad, H.; Saha, T.K.; Poor, H.V.; Wood, K.L. Transforming energy networks via peer-to-peer energy trading: The Potential of game-theoretic approaches. *IEEE Signal Process. Mag.* **2018**, *35*, 90–111. [CrossRef]

9. Woo, J.; Choi, H.; Ahn, J. Well-to-wheel analysis of greenhouse gas emissions for electric vehicles based on electricity generation mix: A global perspective. *Transp. Res. Part D Transp. Environ.* **2017**, *51*, 340–350. [CrossRef]

10. Saber, A.Y.; Venayagamoorthy, G.K. Plug-in vehicles and renewable energy sources for cost and emission reductions. *IEEE Trans. Ind. Electron.* **2010**, *58*, 1229–1238. [CrossRef]

11. Bhatti, A.R.; Salam, Z.; Aziz, M.J.A.; Yee, K.P.; Ashique, R. Electric vehicles charging using photovoltaic: Status and technological review. *Renew. Sustain. Energy Rev.* **2016**, *54*, 34–47. [CrossRef]

12. Hoarau, Q.; Perez, Y. Interactions between electric mobility and photovoltaic generation: A review. *Renew. Sustain. Energy Rev.* **2018**, *94*, 510–522. [CrossRef]

13. Shepero, M.; Munkhammar, J.; Widén, J.; Bishop, J.D.; Boström, T. Modeling of photovoltaic power generation and electric vehicles charging on city-scale: A review. *Renew. Sustain. Energy Rev.* **2018**, *89*, 61–71. [CrossRef]

14. Fachrizal, R.; Shepero, M.; Van Der Meer, D.; Munkhammar, J.; Widén, J. Smart charging of electric vehicles considering photovoltaic power production and electricity consumption: A review. *eTransportation* **2020**, *4*, 100056. [CrossRef]

15. Yong, J.Y.; Ramachandaramurthy, V.K.; Tan, K.M.; Mithulananthan, N. A review on the state-of-the-art technologies of electric vehicle, its impacts and prospects. *Renew. Sustain. Energy Rev.* **2015**, *49*, 365–385. [CrossRef]

16. Ma, C.-T. System planning of grid-connected electric vehicle charging stations and key technologies: A review. *Energies* **2019**, *12*, 4201. [CrossRef]

17. Ahmadian, A.; Mohammadi-Ivatloo, B.; Elkamel, A. A Review on plug-in electric vehicles: Introduction, Current status, and load modeling techniques. *J. Mod. Power Syst. Clean Energy* **2020**, *8*, 412–425. [CrossRef]

18. Richardson, D.B. Electric vehicles and the electric grid: A review of modeling approaches, impacts, and renewable energy integration. *Renew. Sustain. Energy Rev.* **2013**, *19*, 247–254. [CrossRef]

19. Su, J.; Lie, T.; Zamora, R. Modelling of large-scale electric vehicles charging demand: A New Zealand case study. *Electr. Power Syst. Res.* **2019**, *167*, 171–182. [CrossRef]

20. Chung, C.-Y.; Youn, E.; Chynoweth, J.S.; Qiu, C.; Chu, C.-C.; Gadh, R. Safety design for smart Electric Vehicle charging with current and multiplexing control. In Proceedings of the 2013 IEEE International Conference on Smart Grid Communications (SmartGridComm), Vancouver, BC, Canada, 21–24 October 2013; pp. 540–545.

21. Zheng, Y.; Niu, S.; Shang, Y.; Shao, Z.; Jian, L. Integrating plug-in electric vehicles into power grids: A comprehensive review on power interaction mode, scheduling methodology and mathematical foundation. *Renew. Sustain. Energy Rev.* **2019**, *112*, 424–439. [CrossRef]

22. He, Y.; Venkatesh, B.; Guan, L. Optimal scheduling for charging and discharging of electric vehicles. *IEEE Trans. Smart Grid* **2012**, *3*, 1095–1105. [CrossRef]

23. Ahn, C.; Li, C.-T.; Peng, H. Optimal decentralized charging control algorithm for electrified vehicles connected to smart grid. *J. Power Sources* **2011**, *196*, 10369–10379. [CrossRef]

24. Hu, J.; You, S.; Lind, M.; Østergaard, J. Coordinated charging of electric vehicles for congestion prevention in the distribution grid. *IEEE Trans. Smart Grid* **2013**, *5*, 703–711. [CrossRef]

25. Uddin, M.; Romlie, M.; Abdullah, M.F.; Halim, S.A.; Abu Bakar, A.H.; Kwang, T.C. A review on peak load shaving strategies. *Renew. Sustain. Energy Rev.* **2018**, *82*, 3323–3332. [CrossRef]

26. Letendre, S.E.; Kempton, W. The V2G concept: A new model for power? *Public Util. Fortn.* **2002**, *140*, 16–26.

27. Guille, C.; Gross, G. A conceptual framework for the vehicle-to-grid (V2G) implementation. *Energy Policy* **2009**, *37*, 4379–4390. [CrossRef]

28. Tan, K.M.; Ramachandaramurthy, V.K.; Yong, J.Y. Integration of electric vehicles in smart grid: A review on vehicle to grid technologies and optimization techniques. *Renew. Sustain. Energy Rev.* **2016**, *53*, 720–732. [CrossRef]

29. Sortomme, E.; El-Sharkawi, M.A. Optimal charging strategies for unidirectional vehicle-to-grid. *IEEE Trans. Smart Grid* **2010**, *2*, 131–138. [CrossRef]

30. Boynuegri, A.; Uzunoglu, M.; Erdinc, O.; Gokalp, E. A new perspective in grid connection of electric vehicles: Different operating modes for elimination of energy quality problems. *Appl. Energy* **2014**, *132*, 435–451. [CrossRef]

31. Turker, H.; Hably, A.; Bacha, S. Housing peak shaving algorithm (HPSA) with plug-in hybrid electric vehicles (PHEVs): Vehicle-to-Home (V2H) and Vehicle-to-Grid (V2G) concepts. In Proceedings of the 4th International Conference on Power Engineering, Energy and Electrical Drives, Istanbul, Turkey, 13–17 May 2013; pp. 753–759.

32. Feldman, D.; Barbose, G.; Margolis, R.; Wiser, R.; Darghouth, N.; Goodrich, A. *Photovoltaic (PV) Pricing Trends: Historical, Recent, and Near-Term Projections*; National Renewable Energy Laboratory: Golden, CO, USA, 2012.

33. Goli, P.; Shireen, W. PV powered smart charging station for PHEVs. *Renew. Energy* **2014**, *66*, 280–287. [CrossRef]

34. Carli, G.; Williamson, S.S. Technical considerations on power conversion for electric and plug-in hybrid electric vehicle battery charging in photovoltaic installations. *IEEE Trans. Power Electron.* **2013**, *28*, 5784–5792. [CrossRef]

35. Van Der Meer, D.; Mouli, G.R.C.; Morales-Espana, G.; Elizondo, L.R.; Bauer, P. Erratum to energy management system with pv power forecast to optimally charge evs at the workplace. *IEEE Trans. Ind. Inform.* **2018**, *14*, 3298. [CrossRef]

36. Oliveira, D.; De Souza, A.Z.; Delboni, L. Optimal plug-in hybrid electric vehicles recharge in distribution power systems. *Electr. Power Syst. Res.* **2013**, *98*, 77–85. [CrossRef]

37. Ahmad, F.; Alam, M.S.; Shahidehpour, M. Profit maximization of microgrid aggregator under power market environment. *IEEE Syst. J.* **2019**, *13*, 3388–3399. [CrossRef]

38. Sarker, M.R.; Dvorkin, Y.; Ortega-Vazquez, M. Optimal participation of an electric vehicle aggregator in day-ahead energy and reserve markets. *IEEE Trans. Power Syst.* **2015**, *31*, 3506–3515. [CrossRef]

39. Reddy, K.R.; Meikandasivam, S. Load flattening and voltage regulation using plug-in electric vehicle's storage capacity with vehicle prioritization using ANFIS. *IEEE Trans. Sustain. Energy* **2020**, *11*, 260–270. [CrossRef]

40. Ko, K.S.; Han, S.; Sung, D.K. Performance-based settlement of frequency regulation for electric vehicle aggregators. *IEEE Trans. Smart Grid* **2018**, *9*, 866–875. [CrossRef]

41. Zhou, K.; Cai, L. Randomized PHEV charging under distribution grid constraints. *IEEE Trans. Smart Grid* **2014**, *5*, 879–887. [CrossRef]

42. Ma, Z.; Callaway, D.S.; Hiskens, I.A. Decentralized charging control of large populations of plug-in electric vehicles. *IEEE Trans. Control. Syst. Technol.* **2011**, *21*, 67–78. [CrossRef]

43. Unda, I.G.; Papadopoulos, P.; Skarvelis-Kazakos, S.; Cipcigan, L.M.; Jenkins, N.; Zabala, E. Management of electric vehicle battery charging in distribution networks with multi-agent systems. *Electr. Power Syst. Res.* **2014**, *110*, 172–179. [CrossRef]

44. Chaudhari, K.; Kandasamy, N.K.; Krishnan, A.; Ukil, A.; Gooi, H.B. Agent-based aggregated behavior modeling for electric vehicle charging load. *IEEE Trans. Ind. Inform.* **2018**, *15*, 856–868. [CrossRef]

45. Liu, M.; Phanivong, P.K.; Shi, Y.; Callaway, D.S. Decentralized charging control of electric vehicles in residential distribution networks. *IEEE Trans. Control. Syst. Technol.* **2019**, *27*, 266–281. [CrossRef]

46. Torreglosa, J.P.; García-Triviño, P.; Fernández-Ramírez, L.M.; Jurado, F. Decentralized energy management strategy based on predictive controllers for a medium voltage direct current photovoltaic electric vehicle charging station. *Energy Convers. Manag.* **2016**, *108*, 1–13. [CrossRef]

47. Weckx, S.; D'Hulst, R.; Driesen, J. Primary and secondary frequency support by a multi-agent demand control system. *IEEE Trans. Power Syst.* **2014**, *30*, 1394–1404. [CrossRef]

48. Paterakis, N.G.; Erdinc, O.; Bakirtzis, A.G.; Catalão, J.P. Optimal household appliances scheduling under day-ahead pricing and load-shaping demand response strategies. *IEEE Trans. Ind. Inform.* **2015**, *11*, 1509–1519. [CrossRef]

49. Xi, X.; Sioshansi, R. Using price-based signals to control plug-in electric vehicle fleet charging. *IEEE Trans. Smart Grid* **2014**, *5*, 1451–1464. [CrossRef]

50. Pan, J.; Jain, R.; Paul, S.; Vu, T.; Saifullah, A.; Sha, M. A internet of things framework for smart energy in buildings: Designs, prototype, and experiments. *IEEE Internet Things J.* **2015**, *2*, 1. [CrossRef]

51. Hu, J.; Morais, H.; Sousa, T.; Lind, M. Electric vehicle fleet management in smart grids: A review of services, optimization and control aspects. *Renew. Sustain. Energy Rev.* **2016**, *56*, 1207–1226. [CrossRef]

52. Fischer, D.; Harbrecht, A.; Surmann, A.; McKenna, R. Electric vehicles' impacts on residential electric local profiles—A stochastic modelling approach considering socio-economic, behavioural and spatial factors. *Appl. Energy* **2019**, 644–658. [CrossRef]

53. Nunes, P.; Figueiredo, R.; Brito, M.C. The use of parking lots to solar-charge electric vehicles. *Renew. Sustain. Energy Rev.* **2016**, *66*, 679–693. [CrossRef]

54. Prakash, K.; Vaithilingam, C.A.; Rajendran, G.; Vaithilingam, C.A. Design and sizing of mobile solar photovoltaic power plant to support rapid charging for electric vehicles. *Energies* **2019**, *12*, 3579. [CrossRef]

55. Denholm, P.; Kuss, M.; Margolis, R.M. Co-benefits of large scale plug-in hybrid electric vehicle and solar PV deployment. *J. Power Sources* **2013**, *236*, 350–356. [CrossRef]

56. Nunes, P.; Farias, T.L.; Brito, M.C. Day charging electric vehicles with excess solar electricity for a sustainable energy system. *Energy* **2015**, *80*, 263–274. [CrossRef]

57. Tulpule, P.; Marano, V.; Yurkovich, S.; Rizzoni, G. Economic and environmental impacts of a PV powered workplace parking garage charging station. *Appl. Energy* **2013**, *108*, 323–332. [CrossRef]

58. Sarkar, J.; Bhattacharyya, S. Operating characteristics of transcritical CO_2 heat pump for simultaneous water cooling and heating. *Arch. Thermodyn.* **2011**, *33*, 23–40. [CrossRef]

59. Kempton, W.; Tomić, J. Vehicle-to-grid power implementation: From stabilizing the grid to supporting large-scale renewable energy. *J. Power Sources* **2005**, *144*, 280–294. [CrossRef]

60. Moghaddam, Z.; Ahmad, I.; Habibi, D.; Phung, Q.V.; Habibi, D. Smart charging strategy for electric vehicle charging stations. *IEEE Trans. Transp. Electrif.* **2018**, *4*, 76–88. [CrossRef]

61. Mouli, G.R.C.; Kefayati, M.; Baldick, R.; Bauer, P. Integrated PV charging of EV fleet based on energy prices, V2G, and offer of reserves. *IEEE Trans. Smart Grid* **2017**, *10*, 1313–1325. [CrossRef]

62. Ioakimidis, C.S.; Thomas, D.; Rycerski, P.; Genikomsakis, K.N. Peak shaving and valley filling of power consumption profile in non-residential buildings using an electric vehicle parking lot. *Energy* **2018**, *148*, 148–158. [CrossRef]

63. Ivanova, A.; Fernandez, J.A.; Crawford, C.; Sui, P.-C. Coordinated charging of electric vehicles connected to a net-metered PV parking lot. In Proceedings of the 2017 IEEE PES Innovative Smart Grid Technologies Conference Europe (ISGT-Europe), Torino, Italy, 26–29 September 2017; pp. 1–6.

64. Mohamed, A.A.; Salehi, V.; Ma, T.; Mohammed, O.A. Real-time energy management algorithm for plug-in hybrid electric vehicle charging parks involving sustainable energy. *IEEE Trans. Sustain. Energy* **2014**, *5*, 577–586. [CrossRef]

65. Liu, N.; Chen, Q.; Liu, J.; Lu, X.; Li, P.; Lei, J.; Zhang, J. A heuristic operation strategy for commercial building microgrids containing EVs and PV system. *IEEE Trans. Ind. Electron.* **2014**, *62*, 2560–2570. [CrossRef]

66. Zhang, Y.; Cai, L. Dynamic charging scheduling for EV parking lots with photovoltaic power system. *IEEE Access* **2018**, *6*, 56995–57005. [CrossRef]

67. Alam, M.J.E.; Muttaqi, K.M.; Sutanto, D. Effective utilization of available PEV battery capacity for mitigation of solar PV impact and grid support with integrated V2G functionality. *IEEE Trans. Smart Grid* **2015**, *7*, 1562–1571. [CrossRef]

68. Wi, Y.-M.; Lee, J.-U.; Joo, S.-K. Electric vehicle charging method for smart homes/buildings with a photovoltaic system. *IEEE Trans. Consum. Electron.* **2013**, *59*, 323–328. [CrossRef]

69. Wu, X.; Hu, X.; Teng, Y.; Qian, S.; Cheng, R. Optimal integration of a hybrid solar-battery power source into smart home nanogrid with plug-in electric vehicle. *J. Power Sources* **2017**, *363*, 277–283. [CrossRef]

70. Eldeeb, H.H.; Faddel, S.; Mohammed, O.A. Multi-objective optimization technique for the operation of grid tied PV powered EV charging station. *Electr. Power Syst. Res.* **2018**, *164*, 201–211. [CrossRef]

71. Van Der Kam, M.; Van Sark, W. Smart charging of electric vehicles with photovoltaic power and vehicle-to-grid technology in a microgrid; a case study. *Appl. Energy* **2015**, *152*, 20–30. [CrossRef]

72. Bhatti, A.R.; Salam, Z. A rule-based energy management scheme for uninterrupted electric vehicles charging at constant price using photovoltaic-grid system. *Renew. Energy* **2018**, *125*, 384–400. [CrossRef]

73. Barone, G.; Buonomano, A.; Calise, F.; Forzano, C.; Palombo, A. Building to vehicle to building concept toward a novel zero energy paradigm: Modelling and case studies. *Renew. Sustain. Energy Rev.* **2019**, *101*, 625–648. [CrossRef]

74. Ghotge, R.; Snow, Y.; Farahani, S.; Lukszo, Z.; Van Wijk, A.J. Optimized scheduling of EV charging in solar parking lots for local peak reduction under eV demand uncertainty. *Energies* **2020**, *13*, 1275. [CrossRef]

75. Fachrizal, R.; Munkhammar, J. Improved photovoltaic self-consumption in residential buildings with distributed and centralized smart charging of electric vehicles. *Energies* **2020**, *13*, 1153. [CrossRef]

76. Nunes, P.; Farias, T.L.; Brito, M.C. Enabling solar electricity with electric vehicles smart charging. *Energy* **2015**, *87*, 10–20. [CrossRef]

77. Kadar, P.; Varga, A. PhotoVoltaic EV charge station. In Proceedings of the 2013 IEEE 11th International Symposium on Applied Machine Intelligence and Informatics (SAMI), Herl'any, Slovenia, 31 January–2 February 2013; pp. 57–60.

78. Brenna, M.; Dolara, A.; Foiadelli, F.; Leva, S.; Longo, M. Urban scale photovoltaic charging stations for electric vehicles. *IEEE Trans. Sustain. Energy* **2014**, *5*, 1234–1241. [CrossRef]

79. Leou, R.-C.; Su, C.-L.; Lu, C.-N. Stochastic analyses of electric vehicle charging impacts on distribution network. *IEEE Trans. Power Syst.* **2013**, *29*, 1055–1063. [CrossRef]

80. Khodayar, M.E.; Wu, L.; Shahidehpour, M. Hourly Coordination of electric vehicle operation and volatile wind power generation in SCUC. *IEEE Trans. Smart Grid* **2012**, *3*, 1271–1279. [CrossRef]

81. Liu, Z.; Wen, F.; Ledwich, G. Optimal siting and sizing of distributed generators in distribution systems considering uncertainties. *IEEE Trans. Power Deliv.* **2011**, *26*, 2541–2551. [CrossRef]

82. Soares, J.; Borges, N.; Ghazvini, M.A.F.; Vale, Z.; Oliveira, P.M. Scenario generation for electric vehicles' uncertain behavior in a smart city environment. *Energy* **2016**, *111*, 664–675. [CrossRef]

83. Chen, Z.; Xiong, R.; Cao, J. Particle swarm optimization-based optimal power management of plug-in hybrid electric vehicles considering uncertain driving conditions. *Energy* **2016**, *96*, 197–208. [CrossRef]

84. Soroudi, A.; Ehsan, M. A possibilistic–probabilistic tool for evaluating the impact of stochastic renewable and controllable power generation on energy losses in distribution networks—A case study. *Renew. Sustain. Energy Rev.* **2011**, *15*, 794–800. [CrossRef]

85. Ahmad, F.; Alam, M.S.; Shariff, S.M.; Krishnamurthy, M. A Cost-efficient approach to ev charging station integrated community microgrid: A case study of Indian power market. *IEEE Trans. Transp. Electrif.* **2019**, *5*, 200–214. [CrossRef]

86. Shepero, M.; Munkhammar, J. Spatial Markov chain model for electric vehicle charging in cities using geographical information system (GIS) data. *Appl. Energy* **2018**, *231*, 1089–1099. [CrossRef]

87. Gupta, N. Gauss-quadrature-based probabilistic load flow method with voltage-dependent loads including WTGS, PV, and EV charging uncertainties. *IEEE Trans. Ind. Appl.* **2018**, *54*, 6485–6497. [CrossRef]

88. Zhou, B.; Yang, X.; Yang, D.; Yang, Z.; Littler, T.; Li, H. Probabilistic load flow algorithm of distribution networks with distributed generators and electric vehicles integration. *Energies* **2019**, *12*, 4234. [CrossRef]

89. Baringo, L.; Amaro, R.S. A stochastic robust optimization approach for the bidding strategy of an electric vehicle aggregator. *Electr. Power Syst. Res.* **2017**, *146*, 362–370. [CrossRef]

90. Sarker, M.R.; Pandžić, H.; Ortega-Vazquez, M. Optimal Operation and services scheduling for an electric vehicle battery swapping station. *IEEE Trans. Power Syst.* **2014**, *30*, 901–910. [CrossRef]

91. Zhao, J.; Wan, C.; Xu, Z.; Wang, J. Risk-based day-ahead scheduling of electric vehicle aggregator using information gap decision theory. *IEEE Trans. Smart Grid* **2015**, *8*, 1609–1618. [CrossRef]

92. Soroudi, A.; Keane, A. Risk averse energy hub management considering plug-in electric vehicles using information gap decision theory. In *Power Systems*; Springer: Singapore, 2015; Volume 89, pp. 107–127.

93. Aien, M.; Fotuhi-Firuzabad, M.; Rashidinejad, M. Probabilistic optimal power flow in correlated hybrid WindPhotovoltaic power systems. *IEEE Trans. Smart Grid* **2014**, *5*, 130–138. [CrossRef]

94. Al-Dahidi, S.; Ayadi, O.; Alrbai, M.; Adeeb, J. Ensemble approach of optimized artificial neural networks for solar photovoltaic power prediction. *IEEE Access* **2019**, *7*, 81741–81758. [CrossRef]

95. Zhao, Q.; Wang, P.; Goel, L.; Ding, Y. Evaluation of nodal reliability risk in a deregulated power system with photovoltaic power penetration. *IET Gener. Transm. Distrib.* **2013**, *8*, 421–430. [CrossRef]

96. Zhao, J.; Wang, W.; Sheng, C. Industrial prediction intervals with data Uncertainty. In *Information Fusion and Data Science*; Springer: Cham, Switzerland, 2018; pp. 159–222.

97. Wu, X.; Wang, X.; Qu, C. A hierarchical framework for generation scheduling of microgrids. *IEEE Trans. Power Deliv.* **2014**, *29*, 2448–2457. [CrossRef]

98. Ehsan, A.; Yang, Q. Coordinated investment planning of distributed multi-type stochastic generation and battery storage in active distribution networks. *IEEE Trans. Sustain. Energy* **2019**, *10*, 1813–1822. [CrossRef]

99. Wang, R.; Wang, P.; Xiao, G. A robust optimization approach for energy generation scheduling in microgrids. *Energy Convers. Manag.* **2015**, *106*, 597–607. [CrossRef]

100. Sun, Y.; Huang, P.; Huang, G. A multi-criteria system design optimization for net zero energy buildings under uncertainties. *Energy Build.* **2015**, *97*, 196–204. [CrossRef]

101. Koraki, D.; Strunz, K. Wind and solar power integration in electricity markets and distribution networks through service-centric virtual power plants. *IEEE Trans. Power Syst.* **2017**, *33*, 473–485. [CrossRef]

102. Chen, Y.; Wang, Y.; Kirschen, D.S.; Zhang, B. Model-free renewable scenario generation using generative adversarial networks. *IEEE Trans. Power Syst.* **2018**, *33*, 3265–3275. [CrossRef]

103. Ahmed, R.; Sreeram, V.; Mishra, Y.; Arif, M. A review and evaluation of the state-of-the-art in PV solar power forecasting: Techniques and optimization. *Renew. Sustain. Energy Rev.* **2020**, *124*, 109792. [CrossRef]

104. Das, U.K.; Soon, T.; Seyedmahmoudian, M.; Mekhilef, S.; Idris, M.Y.I.; Van Deventer, W.; Horan, B.; Stojcevski, A. Forecasting of photovoltaic power generation and model optimization: A review. *Renew. Sustain. Energy Rev.* **2018**, *81*, 912–928. [CrossRef]

105. Ramadhani, U.H.; Shepero, M.; Munkhammar, J.; Widén, J.; Etherden, N. Review of probabilistic load flow approaches for power distribution systems with photovoltaic generation and electric vehicle charging. *Int. J. Electr. Power Energy Syst.* **2020**, *120*, 106003. [CrossRef]

106. Zakaria, A.; Ismail, F.B.; Lipu, M.S.H.; Hannan, M. Uncertainty models for stochastic optimization in renewable energy applications. *Renew. Energy* **2020**, *145*, 1543–1571. [CrossRef]

107. Ehsan, A.; Yang, Q. State-of-the-art techniques for modelling of uncertainties in active distribution network planning: A review. *Appl. Energy* **2019**, *239*, 1509–1523. [CrossRef]

108. Kumar, K.P.; Saravanan, B. Recent techniques to model uncertainties in power generation from renewable energy sources and loads in microgrids—A review. *Renew. Sustain. Energy Rev.* **2017**, *71*, 348–358. [CrossRef]

109. Kawamura, N.; Muta, M. Development of solar charging system for plug-in hybrid electric vehicles and electric vehicles. In Proceedings of the 2012 International Conference on Renewable Energy Research and Applications (ICRERA), Nagasaki, Japan, 11–14 November 2012; pp. 1–5.

110. Castello, C.C.; LaClair, T.J.; Curt Maxey, L. Control strategies for electric vehicle (EV) charging using renewables and local storage. In Proceedings of the 2014 IEEE Transportation Electrification Conference and Expo (ITEC), Dearborn, MI, USA, 15–18 June 2014; pp. 1–7.

111. Prusty, B.R.; Jena, D. A critical review on probabilistic load flow studies in uncertainty constrained power systems with photovoltaic generation and a new approach. *Renew. Sustain. Energy Rev.* **2017**, *69*, 1286–1302. [CrossRef]

112. Hong, T.; Fan, S. Probabilistic electric load forecasting: A tutorial review. *Int. J. Forecast.* **2016**, *32*, 914–938. [CrossRef]

113. Jordehi, A.R. How to deal with uncertainties in electric power systems? A review. *Renew. Sustain. Energy Rev.* **2018**, *96*, 145–155. [CrossRef]

114. Verbič, G.; Canizares, C.A. Probabilistic optimal power flow in electricity markets based on a two-point estimate method. *IEEE Trans. Power Syst.* **2006**, *21*, 1883–1893. [CrossRef]

115. Morales, J.; Perez-Ruiz, J. Point estimate schemes to solve the probabilistic power flow. *IEEE Trans. Power Syst.* **2007**, *22*, 1594–1601. [CrossRef]

116. Alaee, S.; Hooshmand, R.-A.; Hemmati, R. Stochastic transmission expansion planning incorporating reliability solved using SFLA meta-heuristic optimization technique. *CSEE J. Power Energy Syst.* **2016**, *2*, 79–86. [CrossRef]

117. Cai, D.; Shi, D.; Chen, J. Probabilistic load flow with correlated input random variables using uniform design sampling. *Int. J. Electr. Power Energy Syst.* **2014**, *63*, 105–112. [CrossRef]

118. Soares, J.; Ghazvini, M.A.F.; Vale, Z.; Oliveira, P.M. A multi-objective model for the day-ahead energy resource scheduling of a smart grid with high penetration of sensitive loads. *Appl. Energy* **2016**, *162*, 1074–1088. [CrossRef]

119. Talari, S.; Haghifam, M.-R.; Yazdaninejad, M. Stochastic-based scheduling of the microgrid operation including wind turbines, photovoltaic cells, energy storages and responsive loads. *IET Gener. Transm. Distrib.* **2015**, *9*, 1498–1509. [CrossRef]

120. El Motaleb, A.M.A.; Bekdache, S.K.; Alvarado-Barrios, L. Optimal sizing for a hybrid power system with wind/energy storage based in stochastic environment. *Renew. Sustain. Energy Rev.* **2016**, *59*, 1149–1158. [CrossRef]

121. Allan, R.; Da Silva, A.; Burchett, R. Evaluation methods and accuracy in probabilistic load flow solutions. *IEEE Trans. Power Appar. Syst.* **1981**, 2539–2546. [CrossRef]

122. Schwippe, J.; Krause, O.; Rehtanz, C. Probabilistic load flow calculation based on an enhanced convolution technique. In Proceedings of the 2009 IEEE Bucharest PowerTech, Bucharest, Romania, 28 June–2 July 2009; pp. 1–6.

Energies **2020**, *13*, 4541

123. Munkhammar, J.; Rydén, J.; Widén, J. Characterizing probability density distributions for household electricity load profiles from high-resolution electricity use data. *Appl. Energy* **2014**, *135*, 382–390. [CrossRef]

124. Li, G.; Zhang, X.-P. Comparison between two probabilistic load flow methods for reliability assessment. In Proceedings of the 2009 IEEE Power & Energy Society General Meeting, Calgary, AB, Canada, 26–30 July 2009; pp. 1–7.

MDPI

St. Alban-Anlage 66

4052 Basel

Switzerland

Tel. +41 61 683 77 34

Fax +41 61 302 89 18

www.mdpi.com

Energies Editorial Office

E-mail: energies@mdpi.com

www.mdpi.com/journal/energies

www.ingramcontent.com/pod-product-compliance
Lightning Source LLC
LaVergne TN
LVHW070853170726
843515LV00005B/639